The Complete
U.S. Army Survival Guide to
Medical
Skills, Tactics, and Techniques

The Complete U.S. Army Survival Guide to
Medical
Skills, Tactics, and Techniques

U.S. Army Survival Series

Edited by

Jay McCullough

Skyhorse Publishing

Skyhorse Publishing books may be purchased in bulk at special discounts for sales promotion, corporate gifts, fund-raising, or educational purposes. Special editions can also be created to specifications. For details, contact the Special Sales Department, Skyhorse Publishing, 307 West 36th Street, 11th Floor, New York, NY 10018 or info@skyhorsepublishing.com.

Skyhorse® and Skyhorse Publishing® are registered trademarks of Skyhorse Publishing, Inc.®, a Delaware corporation.

Visit our website at www.skyhorsepublishing.com.

10 9 8 7 6 5 4 3

Library of Congress Cataloging-in-Publication Data is available on file.

Cover design by Tom Lau

Print ISBN: 978-1-5107-0741-2
Ebook ISBN: 978-1-5107-0746-7

Printed in the United States of America

CONTENTS

FOREWORD

Everyone should acquire basic First Aid skills. *The U.S. Army Survival Guide to Medical Skills, Tactics, and Techniques* is as good a place as any for a starting point on learning how to be prepared for many of the dire circumstances that occur on a surprisingly frequent basis. U.S. Army medics have seen and treated as wide a variety of acute medical crises as can be imagined, and the information that makes up this tome is a distillation of their many invaluable manuals and guides. For the casual reader who would like to be reasonably prepared, it would do well to commit much of its contents to memory. For those who would make a serious study of emergency medical treatment, it should be a treasured and interesting counterpoint to civilian emergency medical guides.

In a traumatic injury situation, there is no substitute for rigorous, accredited study and relevant hands-on training. Barring that, a clear head, regular training with relevant materials, and a familiarity with a variety of common medical problems may well save a life. It is my sincere hope that you will never need to employ what you may learn in *The U.S. Army Survival Guide to Medical Skills, Tactics, and Techniques*, but that if you do, that it serves you well.

—Jay McCullough
North Haven, Connecticut
February 2016

INTRODUCTION

Foremost among the many problems that can compromise a survivor's ability to return to safety are medical problems resulting from parachute descent and landing, extreme climates, ground combat, evasion, and illnesses contracted in captivity.

Many evaders and survivors have reported difficulty in treating injuries and illness due to the lack of training and medical supplies. For some, this led to capture or surrender.

Survivors have related feeling of apathy and helplessness because they could not treat themselves in this environment. The ability to treat themselves increased their morale and cohesion and aided in their survival and eventual return to friendly forces.

One man with a fair amount of basic medical knowledge can make a difference in the lives of many. Without qualified medical personnel available, it is you who must know what to do to stay alive.

Part II meets the emergency medical training needs of individual soldiers. Because medical personnel will not always be readily available, the non-medical soldiers will have to rely heavily on their own skills and knowledge of life-sustaining methods to survive on the integrated battlefield. This manual also addresses first aid measures for other life-threatening situations. It outlines both self-treatment (self-aid) and aid to other soldiers (buddy aid). More importantly, this manual emphasizes prompt and effective action in sustaining life and preventing or minimizing further suffering. First aid is the emergency care given to the sick, injured, or wounded before being treated by medical personnel. The Army Dictionary defines first aid as "urgent and immediate life saving and other measures which can be performed for casualties by non-medical personnel when medical personnel are not immediately available." Non-medical soldiers have received basic first aid training and should remain skilled in the correct procedures for giving first aid. A combat lifesaver is a non-medical soldier who has been trained to provide emergency care. This includes administering intravenous infusions to casualties as his combat mission permits. Normally, each squad, team, or crew will have one member who is a combat lifesaver. This manual is directed to all soldiers. The procedures discussed apply to all types of casualties and the measures described are for use by both male and female soldiers.

Part II has been designed to provide a ready reference for the individual soldier on first aid. Only the information necessary to support and sustain proficiency in first aid has been boxed and the task number has been listed. In addition, these first aid tasks for Skill Level 1 have been listed in Appendix G. The task number, title, and specific paragraph of the appropriate information is provided in the event a cross-reference is desired.

Commercial products (trade names or trademarks) mentioned in this publication are to provide descriptive information and for illustrative purposes only. Their use does not imply endorsement by the Department of Defense.

REQUIREMENTS FOR MAINTENANCE OF HEALTH: OVERVIEW

To survive, you need water and food. You must also have and apply highpersonal hygiene standards.

Water
Your body loses water through normal body processes (sweating, urinating, and defecating). During average daily exertion when the atmospheric temperature is 20 degrees Celsius (C) (68 degrees Fahrenheit), the average adult loses and therefore requires 2 to 3 liters of water daily. Other factors, such as heat exposure, cold exposure, intense activity, high altitude, burns, or illness, can cause your body to lose more water. You must replace this water.

Dehydration results from inadequate replacement of lost body fluids. It decreases your efficiency and, if injured, increases your susceptibility to severe shock. Consider the following results of body fluid loss:

- A 5 percent loss of body fluids results in thirst, irritability, nausea, and weakness.
- A 10 percent loss results in dizziness, headache, inability to walk, and a tingling sensation in the limbs.
- A 15 percent loss results in dim vision, painful urination, swollen tongue, deafness, and a numb feeling in the skin.
- A loss greater than 15 percent of body fluids may result in death.

The most common signs and symptoms of dehydration are—

- Dark urine with a very strong odor.
- Low urine output.
- Dark, sunken eyes.
- Fatigue.
- Emotional instability.
- Loss of skin elasticity.
- Delayed capillary refill in fingernail beds.
- Trench line down center of tongue.
- Thirst. Last on the list because you are already 2 percent dehydrated by the time you crave fluids.

You replace the water as you lose it. Trying to make up a deficit is difficult in a survival situation, and thirst is not a sign of how much water you need.

Most people cannot comfortably drink more than 1 liter of water at a time. So, even when not thirsty, drink small amounts of water at regular intervals each hour to prevent dehydration.

If you are under physical and mental stress or subject to severe conditions, increase your water intake. Drink enough liquids to maintain a urine output of at least 0.5 liter every 24 hours.

In any situation where food intake is low, drink 6 to 8 liters of water per day. In an extreme climate, especially an arid one, the average person can lose 2.5 to 3.5 liters of water per hour. In this type of climate, you should drink 14 to 30 liters of water per day.

With the loss of water there is also a loss of electrolytes (body salts). The average diet can usually keep up with these losses but in an extreme situation or illness, additional sources need to be provided. A mixture of 0.25 teaspoon of salt to 1 liter of water will provide a concentration that the body tissues can readily absorb.

Of all the physical problems encountered in a survival situation, the loss of water is the most preventable. The following are basic guidelines for the prevention of dehydration:

- *Always drink water when eating.* Water is used and consumed as a part of the digestion process and can lead to dehydration.
- *Acclimatize.* The body performs more efficiently in extreme conditions when acclimatized.
- *Conserve sweat not water.* Limit sweat-producing activities but drink water.
- *Ration water.* Until you find a suitable source, ration your water sensibly. A daily intake of 500 cubic centimeter (0.5 liter) of a sugar-water mixture (2 teaspoons per liter) will suffice to prevent severe dehydration for at least a week, provided you keep water losses to a minimum by limiting activity and heat gain or loss.

You can estimate fluid loss by several means. A standard field dressing holds about 0.25 liter (one-fourth canteen) of blood. A soaked T-shirt holds 0.5 to 0.75 liter.

You can also use the pulse and breathing rate to estimate fluid loss. Use the following as a guide:

- With a 0.75 liter loss the wrist pulse rate will be under 100 beats per minute and the breathing rate 12 to 20 breaths per minute.
- With a 0.75 to 1.5 liter loss the pulse rate will be 100 to 120 beats per minute and 20 to 30 breaths per minute.
- With a 1.5 to 2 liter loss the pulse rate will be 120 to 140 beats per minute and 30 to 40 breaths per minute. Vital signs above these rates require more advanced care.

Food

Although you can live several weeks without food, you need an adequate amount to stay healthy. Without food your mental and physical capabilities will deteriorate rapidly, and you will become weak. Food replenishes the substances that your body burns and provides energy. It provides vitamins, minerals, salts, and other elements essential to good health. Possibly more important, it helps morale.

The two basic sources of food are plants and animals (including fish). In varying degrees both provide the calories, carbohydrates, fats, and proteins needed for normal daily body functions.

Calories are a measure of heat and potential energy. The average person needs 2,000 calories per day to function at a minimum level. An adequate amount of carbohydrates, fats, and proteins without an adequate caloric intake will lead to starvation and cannibalism of the body's own tissue for energy.

Plant Foods

These foods provide carbohydrates—the main source of energy. Many plants provide enough protein to keep the body at normal efficiency. Although plants may not provide a balanced diet, they will sustain you even in the arctic, where meat's heat-producing qualities are normally essential. Many plant foods such as nuts and seeds will give you enough protein and oils for normal efficiency. Roots, green vegetables, and plant food containing natural sugar will provide calories and carbohydrates that give the body natural energy.

The food value of plants becomes more and more important if you are eluding the enemy or if you are in an area where wildlife is scarce. For instance—

- You can dry plants by wind, air, sun, or fire. This retards spoilage so that you can store or carry the plant food with you to use when needed.
- You can obtain plants more easily and more quietly than meat. This is extremely important when the enemy is near.

Animal Foods

Meat is more nourishing than plant food. In fact, it may even be more readily available in some places. However, to get meat, you need to know the habits of, and how to capture, the various wildlife.

To satisfy your immediate food needs, first seek the more abundant and more easily obtained wildlife, such as insects, crustaceans, mollusks, fish, and reptiles. These can satisfy your immediate hunger while you are preparing traps and snares for larger game.

Personal Hygiene

In any situation, cleanliness is an important factor in preventing infection and disease. It becomes even more important in a survival situation. Poor hygiene can reduce your chances of survival.

A daily shower with hot water and soap is ideal, but you can stay clean without this luxury. Use a cloth and soapy water to wash yourself. Pay special attention to the feet, armpits, crotch, hands, and hair as these are prime areas for infestation and infection. If water is scarce, take an "air" bath. Remove as much of your clothing as practical and expose your body to the sun and air for at least 1 hour. Be careful not to sunburn.

If you don't have soap, use ashes or sand, or make soap from animal fat and wood ashes, if your situation allows. To make soap—

- Extract grease from animal fat by cutting the fat into small pieces and cooking them in a pot.
- Add enough water to the pot to keep the fat from sticking as it cooks.
- Cook the fat slowly, stirring frequently.
- After the fat is rendered, pour the grease into a container to harden.
- Place ashes in a container with a spout near the bottom.
- Pour water over the ashes and collect the liquid that drips out of the spout in a separate container. This liquid is the potash or lye.
- Another way to get the lye is to pour the slurry (the mixture of ashes and water) through a straining cloth.
- In a cooking pot, mix two parts grease to one part potash.
- Place this mixture over a fire and boil it until it thickens.

After the mixture—the soap—cools, you can use it in the semiliquid state directly from the pot. You can also pour it into a pan, allow it to harden, and cut it into bars for later use.

Keep Your Hands Clean
Germs on your hands can infect food and wounds. Wash your hands after handling any material that is likely to carry germs, after visiting the latrine, after caring for the sick, and before handling any food, food utensils, or drinking water. Keep your fingernails closely trimmed and clean, and keep your fingers out of your mouth.

Keep Your Hair Clean
Your hair can become a haven for bacteria or fleas, lice, and other parasites.

Keeping your hair clean, combed, and trimmed helps you avoid this danger.

Keep Your Clothing Clean

Keep your clothing and bedding as clean as possible to reduce the chance of skin infection as well as to decrease the danger of parasitic infestation. Clean your outer clothing whenever it becomes soiled. Wear clean underclothing and socks each day. If water is scarce, "air" clean your clothing by shaking, airing, and sunning it for 2 hours. If you are using a sleeping bag, turn it inside out after each use, fluff it, and air it.

Keep Your Teeth Clean

Thoroughly clean your mouth and teeth with a toothbrush at least once each day. If you don't have a toothbrush, make a chewing stick. Find a twig about 20 centimeters long and 1 centimeter wide. Chew one end of the stick to separate the fibers. Now brush your teeth thoroughly. Another way is to wrap a clean strip of cloth around your fingers and rub your teeth with it to wipe away food particles. You can also brush your teeth with small amounts of sand, baking soda, salt, or soap. Then rinse your mouth with water, salt water, or willow bark tea. Also, flossing your teeth with string or fiber helps oral hygiene.

If you have cavities, you can make temporary fillings by placing candle wax, tobacco, aspirin, hot pepper, toothpaste or powder, or portions of a ginger root into the cavity. Make sure you clean the cavity by rinsing or picking the particles out of the cavity before placing a filling in the cavity.

Take Care of Your Feet

To prevent serious foot problems, break in your shoes before wearing them on any mission. Wash and massage your feet daily. Trim your toenails straight across. Wear an insole and the proper size of dry socks. Powder and check your feet daily for blisters.

If you get a small blister, do not open it. An intact blister is safe from infection. Apply a padding material around the blister to relieve pressure and reduce friction. If the blister bursts, treat it as an open wound. Clean and dress it daily and pad around it. Leave large blisters intact. To avoid having the blister burst or tear under pressure and cause a painful and open sore, do the following:

- Obtain a sewing-type needle and a clean or sterilized thread.
- Run the needle and thread through the blister after cleaning the blister.
- Detach the needle and leave both ends of the thread hanging out of the blister. The thread will absorb the liquid inside.

This reduces the size of the hole and ensures that the hole does not close up.
- Pad around the blister.

Get Sufficient Rest

You need a certain amount of rest to keep going. Plan for regular rest periods of at least 10 minutes per hour during your daily activities. Learn to make yourself comfortable under less than ideal conditions. A change from mental to physical activity or vice versa can be refreshing when time or situation does not permit total relaxation.

Keep Camp Site Clean

Do not soil the ground in the camp site area with urine or feces. Use latrines, if available. When latrines are not available, dig "cat holes" and cover the waste. Collect drinking water upstream from the camp site. Purify all water.

MEDICAL EMERGENCIES

Medical problems and emergencies you may be faced with include breathing problems, severe bleeding, and shock.

Breathing Problems

Any one of the following can cause airway obstruction, resulting in stopped breathing

- Foreign matter in mouth of throat that obstructs the opening to the trachea.
- Face or neck injuries.
- Inflammation and swelling of mouth and throat caused by inhaling smoke, flames, and irritating vapors or by an allergic reaction.
- "Kink" in the throat (caused by the neck bent forward so that the chin rests upon the chest) may block the passage of air.
- Tongue blocks passage of air to the lungs upon unconsciousness.
- When an individual is unconscious, the muscles of the lower jaw and tongue relax as the neck drops forward, causing the lower jaw to sag and the tongue to drop back and block the passage of air.

Severe Bleeding

Severe bleeding from any major blood vessel in the body is extremely dangerous. The loss of 1 liter of blood will produce moderate symptoms of shock. The loss of 2 liters will produce a severe state of shock that places the body in extreme danger. The loss of 3 liters is usually fatal.

Shock

Shock (acute stress reaction) is not a disease in itself. It is a clinical condition characterized by symptoms that arise when cardiac output is insufficient to fill the arteries with blood under enough pressure to provide an adequate blood supply to the organs and tissues.

LIFESAVING STEPS

Control panic, both your own and the victim's. Reassure him and try to keep him quiet.

Perform a rapid physical exam. Look for the cause of the injury and follow the ABCs of first aid, starting with the airway and breathing, but be discerning. A person may die from arterial bleeding more quickly than from an airway obstruction in some cases.

Open Airway and Maintain

You can open an airway and maintain it by using the following steps.

Step 1. Check if the victim has a partial or complete airway obstruction. If he can cough or speak, allow him to clear the obstruction naturally. Stand by, reassure the victim, and be ready to clear his airway and perform mouth-to-mouth resuscitation should he become unconscious. If his airway is completely obstructed, administer abdominal thrusts until the obstruction is cleared.

Step 2. Using a finger, quickly sweep the victim's mouth clear of any foreign objects, broken teeth, dentures, sand.

Step 3. Using the jaw thrust method, grasp the angles of the victim's lower jaw and lift with both hands, one on each side, moving the jaw forward. For stability, rest your elbows on the surface on which the victim is lying. If his lips are closed, gently open the lower lip with your thumb (Figure I-1).

Step 4. With the victim's airway open, pinch his nose closed with your thumb and forefinger and blow two complete breaths into his lungs. Allow the lungs to deflate after the second inflation and perform the following:

- Look for his chest to rise and fall.
- Listen for escaping air during exhalation.
- Feel for flow of air on your cheek.

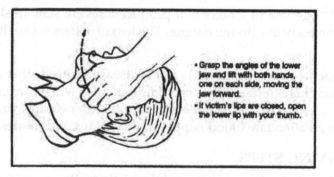

• Grasp the angles of the lower jaw and lift with both hands, one on each side, moving the jaw forward.
• If victim's lips are closed, open the lower lip with your thumb.

Figure I-1: Jaw thrust method

Step 5. If the forced breaths do not stimulate spontaneous breathing, maintain the victim's breathing by performing mouth-to-mouth resuscitation.

Step 6. There is danger of the victim vomiting during mouth-to-mouth resuscitation. Check the victim's mouth periodically for vomit and clear as needed.

Note: Cardiopulmonary resuscitation (CPR) may be necessary after cleaning the airway, but only after major bleeding is under control.

Control Bleeding
In a survival situation, you must control serious bleeding immediately because replacement fluids normally are not available and the victim can die within a matter of minutes. External bleeding falls into the following classifications (according to its source):

- *Arterial.* Blood vessels called arteries carry blood away from the heart and through the body. A cut artery issues bright red blood from the wound in distinct spurts or pulses that correspond to the rhythm of the heartbeat. Because the blood in the arteries is under high pressure, an individual can lose a large volume of blood in a short period when damage to an artery of significant size occurs. Therefore, arterial bleeding is the most serious type of bleeding. If not controlled promptly, it can be fatal.
- *Venous.* Venous blood is blood that is returning to the heart through blood vessels called veins. A steady flow of dark red, maroon, or bluish blood characterizes bleeding from a vein. You can usually control venous bleeding more easily than arterial bleeding.

- *Capillary*. The capillaries are the extremely small vessels that connect the arteries with the veins. Capillary bleeding most commonly occurs in minor cuts and scrapes. This type of bleeding is not difficult to control.

You can control external bleeding by direct pressure, indirect (pressure points) pressure, elevation, digital ligation, or tourniquet.

Direct Pressure

The most effective way to control external bleeding is by applying pressure directly over the wound. This pressure must not only be firm enough to stop the bleeding, but it must also be maintained long enough to "seal off" the damaged surface.

If bleeding continues after having applied direct pressure for 30 minutes, apply a pressure dressing. This dressing consists of a thick dressing of gauze or other suitable material applied directly over the wound and held in place with a tightly wrapped bandage (Figure I-2).

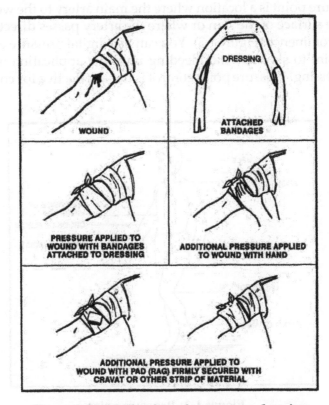

Figure I-2: Application of a pressure dressing

It should be tighter than an ordinary compression bandage but not so tight that it impairs circulation to the rest of the limb. Once you apply the dressing, do not remove it, even when the dressing becomes blood soaked.

Leave the pressure dressing in place for 1 or 2 days, after which you can remove and replace it with a smaller dressing.

In the long-term survival environment, make fresh, daily dressing changes and inspect for signs of infection.

Elevation

Raising an injured extremity as high as possible above the heart's level slows blood loss by aiding the return of blood to the heart and lowering the blood pressure at the wound. However, elevation alone will not control bleeding entirely; you must also apply direct pressure over the wound. When treating a snakebite, however, keep the extremity lower than the heart.

Pressure Points

A pressure point is a location where the main artery to the wound lies near the surface of the skin or where the artery passes directly over a bony prominence (Figure I-3). You can use digital pressure on a pressure point to slow arterial bleeding until the application of a pressure dressing. Pressure point control is not as effective for controlling

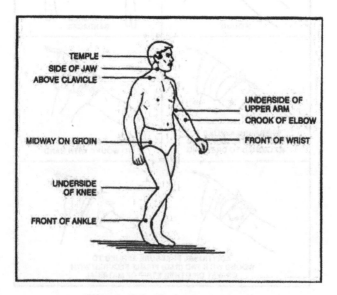

Figure I-3: Pressure points

bleeding as direct pressure exerted on the wound. It is rare when a single major compressible artery supplies a damaged vessel.

If you cannot remember the exact location of the pressure points, follow this rule: Apply pressure at the end of the joint just above the injured area. On hands, feet, and head, this will be the wrist, ankle, and neck respectively.

> **WARNING**
> Use caution when applying pressure to the neck. Too much pressure for too long may cause unconsciousness or death. Never place a tourniquet around the neck.

Maintain pressure points by placing a round stick in the joint, bending the joint over the stick, and then keeping it tightly bent by lashing. By using this method to maintain pressure, it frees your hands to work in other areas.

Digital Ligation
You can stop major bleeding immediately or slow it down by applying pressure with a finger or two on the bleeding end of the vein or artery. Maintain the pressure until the bleeding stops or slows down enough to apply a pressure bandage, elevation, and so forth.

Tourniquet
Use a tourniquet only when direct pressure over the bleeding point and all other methods did not control the bleeding. If you leave a tourniquet in place too long, the damage to the tissues can progress to gangrene, with a loss of the limb later. An improperly applied tourniquet can also cause permanent damage to nerves and other tissues at the site of the constriction.

If you must use a tourniquet, place it around the extremity, between the wound and the heart, 5 to 10 centimeters above the wound site (Figure I-4). Never place it directly over the wound or a fracture. Use a stick as a handle to tighten the tourniquet and tighten it only enough to stop blood flow. When you have tightened the tourniquet, bind the free end of the stick to the limb to prevent unwinding.

After you secure the tourniquet, clean and bandage the wound. A lone survivor does not remove or release an applied tourniquet. In a buddy system, however, the buddy can release the tourniquet

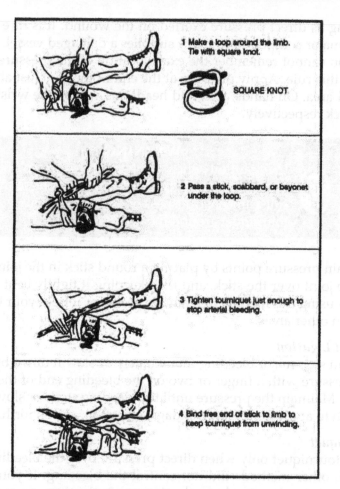

Figure I-4: Application of a tourniquet

pressure every 10 to 15 minutes for 1 or 2 minutes to let blood flow to the rest of the extremity to prevent limb loss.

Prevent and Treat Shock

Anticipate shock in all injured personnel. Treat all injured persons as follows, regardless of what symptoms appear (Figure I-5):

- If the victim is conscious, place him on a level surface with the lower extremities elevated 15 to 20 centimeters.
- If the victim is unconscious, place him on his side or abdomen with his head turned to one side to prevent choking on vomit, blood, or other fluids.

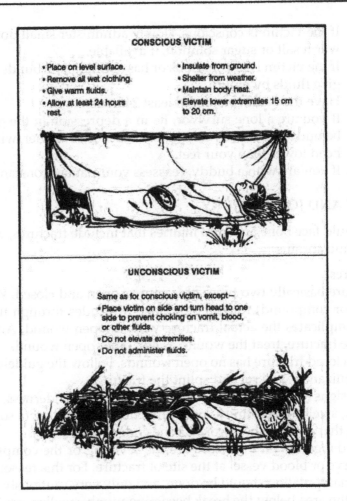

CONSCIOUS VICTIM

- Place on level surface.
- Remove all wet clothing.
- Give warm fluids.
- Allow at least 24 hours rest.
- Insulate from ground.
- Shelter from weather.
- Maintain body heat.
- Elevate lower extremities 15 cm to 20 cm.

UNCONSCIOUS VICTIM

Same as for conscious victim, except—
- Place victim on side and turn head to one side to prevent choking on vomit, blood, or other fluids.
- Do not elevate extremities.
- Do not administer fluids.

Figure I-5: Treatment for shock

- If you are unsure of the best position, place the victim perfectly flat.
- Once the victim is in a shock position, do not move him.
- Maintain body heat by insulating the victim from the surroundings and, in some instances, applying external heat.
- If wet, remove all the victim's wet clothing as soon as possible and replace with dry clothing.
- Improvise a shelter to insulate the victim from the weather.
- Use warm liquids or foods, a pre-warmed sleeping bag, another person, warmed water in canteens, hot rocks wrapped in clothing, or fires on either side of the victim to provide external warmth.

- If the victim is conscious, slowly administer small doses of a warm salt or sugar solution, if available.
- If the victim is unconscious or has abdominal wounds, do not give fluids by mouth.
- Have the victim rest for at least 24 hours.
- If you are a lone survivor, lie in a depression in the ground, behind a tree, or any other place out of the weather, with your head lower than your feet.
- If you are with a buddy, reassess your patient constantly.

BONE AND JOINT INJURY

You could face bone and joint injuries that include fractures, dislocations, and sprains.

Fractures

There are basically two types of fractures: open and closed. With an open (or compound) fracture, the bone protrudes through the skin and complicates the actual fracture with an open wound. After setting the fracture, treat the wound as any other open wound.

The closed fracture has no open wounds. Follow the guidelines for immobilization, and set and splint the fracture.

The signs and symptoms of a fracture are pain, tenderness, discoloration, swelling deformity, loss of function, and grating (a sound or feeling that occurs when broken bone ends rub together).

The dangers with a fracture are the severing or the compression of a nerve or blood vessel at the site of fracture. For this reason minimum manipulation should be done, and only very cautiously. If you notice the area below the break becoming numb, swollen, cool to the touch, or turning pale, and the victim shows signs of shock, a major vessel may have been severed. You must control this internal bleeding. Rest the victim for shock, and replace lost fluids.

Often you must maintain traction during the splinting and healing process.

You can effectively pull smaller bones such as the arm or lower leg by hand. You can create traction by wedging a hand or foot in the V-notch of a tree and pushing against the tree with the other extremity. You can then splint the break.

Very strong muscles hold a broken thighbone (femur) in place making it difficult to maintain traction during healing. You can make an improvised traction splint using natural material (Figure I-6) as follows:

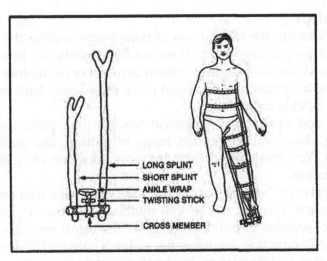

Figure I-6: Improvised traction splint

- Get two forked branches or saplings at least 5 centimeters in diameter. Measure one from the patient's armpit to 20 to 30 centimeters past his unbroken leg. Measure the other from the groin to 20 to 30 centimeters past the unbroken leg. Ensure that both extend an equal distance beyond the end of the leg.
- Pad the two splints. Notch the ends without forks and lash a 20- to 30-centimeter cross member made from a 5-centimeter diameter branch between them.
- Using available material (vines, cloth, rawhide), tie the splint around the upper portion of the body and down the length of the broken leg. Follow the splinting guidelines.
- With available material, fashion a wrap that will extend around the ankle, with the two free ends tied to the cross member.
- Place a 10- by 2.5-centimeter stick in the middle of the free ends of the ankle wrap between the cross member and the foot. Using the stick, twist the material to make the traction easier.
- Continue twisting until the broken leg is as long or slightly longer than the unbroken leg.
- Lash the stick to maintain traction.

Note: Over time you may lose traction because the material weakened. Check the traction periodically. If you must change or repair the splint, maintain the traction manually for a short time.

Dislocations

Dislocations are the separations of bone joints causing the bones to go out of proper alignment. These misalignments can be extremely painful and can cause an impairment of nerve or circulatory function below the area affected. You must place these joints back into alignment as quickly as possible.

Signs and symptoms of dislocations are joint pain, tenderness, swelling, discoloration, limited range of motion, and deformity of the joint. You treat dislocations by reduction, immobilization, and rehabilitation.

Reduction or "setting" is placing the bones back into their proper alignment. You can use several methods, but manual traction or the use of weights to pull the bones are the safest and easiest. Once performed, reduction decreases the victim's pain and allows for normal function and circulation. Without an X ray, you can judge proper alignment by the look and feel of the joint and by comparing it to the joint on the opposite side.

Immobilization is nothing more than splinting the dislocation after reduction. You can use any field-expedient material for a splint or you can splint an extremity to the body. The basic guidelines for splinting are—

- Splint above and below the fracture site.
- Pad splints to reduce discomfort.
- Check circulation below the fracture after making each tie on the splint.

To rehabilitate the dislocation, remove the splints after 7 to 14 days. Gradually use the injured joint until fully healed.

Sprains

The accidental overstretching of a tendon or ligament causes sprains. The signs and symptoms are pain, swelling, tenderness, and discoloration (black and blue).

When treating sprains, think RICE—

R - Rest injured area.

I - Ice for 24 hours, then heat after that.

C - Compression-wrapping and/or splinting to help stabilize. If possible, leave the boot on a sprained ankle unless circulation is compromised.

E - Elevation of the affected area.

BITES AND STINGS

Insects and related pests are hazards in a survival situation. They not only cause irritations, but they are often carriers of diseases that cause severe allergic reactions in some individuals. In many parts of the world you will be exposed to serious, even fatal, diseases not encountered in the United States.

Ticks can carry and transmit diseases, such as Rocky Mountain spotted fever common in many parts of the United States. Ticks also transmit the Lyme disease.

Mosquitoes may carry malaria, dengue, and many other diseases.

Flies can spread disease from contact with infectious sources. They are causes of sleeping sickness, typhoid, cholera, and dysentery.

Fleas can transmit plague.

Lice can transmit typhus and relapsing fever.

The best way to avoid the complications of insect bites and stings is to keep immunizations (including booster shots) up-to-date, avoid insect-infested areas, use netting and insect repellent, and wear all clothing properly.

If you get bitten or stung, do not scratch the bite or sting, it might become infected. Inspect your body at least once a day to ensure there are no insects attached to you. If you find ticks attached to your body, cover them with a substance, such as Vaseline, heavy oil, or tree sap, that will cut off their air supply. Without air, the tick releases its hold, and you can remove it. Take care to remove the whole tick. Use tweezers if you have them. Grasp the tick where the mouth parts are attached to the skin. Do not squeeze the tick's body. Wash your hands after touching the tick. Clean the tick wound daily until healed.

Treatment

It is impossible to list the treatment of all the different types of bites and stings. Threat bites and stings as follows:

- If antibiotics are available for your use, become familiar with them before deployment and use them.
- Predeployment immunizations can prevent most of the common diseases carried by mosquitoes and some carried by flies.
- The common fly-borne diseases are usually treatable with penicillins or erythromycin.

- Most tick-, flea-, louse-, and mite-borne diseases are treatable with tetracycline.
- Most antibiotics come in 250 milligram (mg) or 500 mg tablets. If you cannot remember the exact dose rate to treat a disease, 2 tablets, 4 times a day for 10 to 14 days will usually kill any bacteria.

Bee and Wasp Stings

If stung by a bee, immediately remove the stinger and venom sac, if attached, by scraping with a fingernail or a knife blade. Do not squeeze or grasp the stinger or venom sac, as squeezing will force more venom into the wound. Wash the sting site thoroughly with soap and water to lessen the chance of a secondary infection.

If you know or suspect that you are allergic to insect stings, always carry an insect sting kit with you.

Relieve the itching and discomfort caused by insect bites by applying—

- Cold compresses.
- A cooling paste of mud and ashes.
- Sap from dandelions.
- Coconut meat.
- Crushed cloves of garlic.
- Onion.

Spider Bites and Scorpion Stings

The black widow spider is identified by a red hourglass on its abdomen. Only the female bites, and it has a neurotoxic venom. The initial pain is not severe, but severe local pain rapidly develops. The pain gradually spreads over the entire body and settles in the abdomen and legs. Abdominal cramps and progressive nausea, vomiting, and a rash may occur. Weakness, tremors, sweating, and salivation may occur. Anaphylactic reactions can occur. Symptoms begin to regress after several hours and are usually gone in a few days. Threat for shock. Be ready to perform CPR. Clean and dress the bite area to reduce the risk of infection. An antivenom is available.

The funnel web spider is a large brown or gray spider found in Australia. The symptoms and the treatment for its bite are as for the black widow spider.

The brown house spider or brown recluse spider is a small, light brown spider identified by a dark brown violin on its back. There is no pain, or so little pain, that usually a victim is not aware of the bite. Within a few hours a painful red area with a mottled cyanotic center appears. Necrosis does not occur in all bites, but usually in 3 to 4 days, a star-shaped, firm area of deep purple discoloration appears at the bite site. The area turns dark and mummified in a week or two. The margins separate and the scab falls off, leaving an open ulcer. Secondary infection and regional swollen lymph glands usually become visible at this stage. The outstanding characteristic of the brown recluse bite is an ulcer that does not heal but persists for weeks or months. In addition to the ulcer, there is often a systemic reaction that is serious and may lead to death.

Reactions (fever, chills, joint pain, vomiting, and a generalized rash occur chiefly in children or debilitated persons.

Tarantulas are large, hairy spiders found mainly in the tropics. Most do not inject venom, but some South American species do. They have large fangs. If bitten, pain and bleeding are certain, and infection is likely. Treat a tarantula bite as for any open wound, and try to prevent infection. If symptoms of poisoning appear, treat as for the bite of the black widow spider.

Scorpions are all poisonous to a greater or lesser degree. There are two different reactions, depending on the species:

- Severe local reaction only, with pain and swelling around the area of the sting. Possible prickly sensation around the mouth and a thick-feeling tongue.
- Severe systemic reaction, with little or no visible local reaction. Local pain may be present. Systemic reaction includes respiratory difficulties, thick-feeling tongue, body spasms, drooling, gastric distention, double vision, blindness, involuntary rapid movement of the eyeballs, involuntary urination and defecation, and heart failure. Death is rare, occurring mainly in children and adults with high blood pressure or illnesses.

Treat scorpion stings as you would a black widow bite.

Snakebites
The chance of a snakebite in a survival situation is rather small, if you are familiar with the various types of snakes and their habitats. However, it could happen and you should know how to treat a snakebite. Deaths from snakebites are rare. More than one-half of the snakebite victims have little or no poisoning, and only about one-quarter

develop serious systemic poisoning. However, the chance of a snake-bite in a survival situation can affect morale, and failure to take preventive measures or failure to treat a snakebite properly can result in needless tragedy.

The primary concern in the treatment of snakebite is to limit the amount of eventual tissue destruction around the bite area.

A bite wound, regardless of the type of animal that inflicted it, can become infected from bacteria in the animal's mouth. With nonpoisonous as well as poisonous snakebites, this local infection is responsible for a large part of the residual damage that results.

Snake venoms not only contain poisons that attack the victim's central nervous system (neurotoxins) and blood circulation (hemotoxins), but also digestive enzymes (cytotoxins) to aid in digesting their prey. These poisons can cause a very large area of tissue death, leaving a large open wound. This condition could lead to the need for eventual amputation if not treated.

Shock and panic in a person bitten by a snake can also affect the person's recovery. Excitement, hysteria, and panic can speed up the circulation, causing the body to absorb the toxin quickly. Signs of shock occur within the first 30 minutes after the bite.

Before you start treating a snakebite, determine whether the snake was poisonous or nonpoisonous. Bites from a nonpoisonous snake will show rows of teeth. Bites from a poisonous snake may have rows of teeth showing, but will have one or more distinctive puncture marks caused by fang penetration. Symptoms of a poisonous bite may be spontaneous bleeding from the nose and anus, blood in the urine, pain at the site of the bite, and swelling at the site of the bite within a few minutes or up to 2 hours later.

Breathing difficulty, paralysis, weakness, twitching, and numbness are also signs of neurotoxic venoms. These signs usually appear 1.5 to 2 hours after the bite.

If you determine that a poisonous snake bit an individual, take the following steps:

- Reassure the victim and keep him still.
- Set up for shock and force fluids or give an intravenous (IV).
- Remove watches, rings, bracelets, or other constricting items.
- Clean the bite area.
- Maintain an airway (especially if bitten near the face or neck) and be prepared to administer mouth-to-mouth resuscitation or CPR.

- Use a constricting band between the wound and the heart.
- Immobilize the site.
- Remove the poison as soon as possible by using a mechanical suction device or by squeezing.

Do not–

- Give the victim alcoholic beverages or tobacco products.
- Give morphine or other central nervous system (CNS) depressors.
- Make any deep cuts at the bite site. Cutting opens capillaries that
- Give morphine or other central nervous system (CNS) depressors.
- Make any deep cuts at the bite site. Cutting opens capillaries that in turn open a direct route into the blood stream for venom and infection.

Note: If medical treatment is over one hour away, make an incision(no longer than 6 millimeters and no deeper than 3 millimeter) over each puncture, cutting just deep enough to enlarge the fang opening, but only through the first or second layer of skin. Place a suction cup over the bite so that you have a good vacuum seal. Suction the bite site 3 to 4 times. Use mouth suction only as a last resort and only if you do not have open sores in your mouth. Spit the envenomed blood out and rinse your mouth with water. This method will draw out 25 to 30 percent of the venom.

- Put your hands on your face or rub your eyes, as venom may be on your hands. Venom may cause blindness.
- Break open the large blisters that form around the bite site.

After caring for the victim as described above, take the following actions to minimize local effects:

- If infection appears, keep the wound open and clean.
- Use heat after 24 to 48 hours to help prevent the spread of local infection. Heat also helps to draw out an infection.
- Keep the wound covered with a dry, sterile dressing.
- Have the victim drink large amounts of fluids until the infection is gone.

WOUNDS

An interruption of the skin's integrity characterizes wounds. These wounds could be open wounds, skin diseases, frostbite, trench foot, and burns.

Open Wounds

Open wounds are serious in a survival situation, not only because of tissue damage and blood loss, but also because they may become infected. Bacteria on the object that made the wound, on the individual's skin and clothing, or on other foreign material or dirt that touches the wound may cause infection.

By taking proper care of the wound you can reduce further contamination and promote healing. Clean the wound as soon as possible after it occurs by—

- Removing or cutting clothing away from the wound.
- Always looking for an exit wound if a sharp object, gun shot, or projectile caused a wound.
- Thoroughly cleaning the skin around the wound.
- Rinsing (not scrubbing) the wound with large amounts of water under pressure. You can use fresh urine if water is not available.

The "open treatment" method is the safest way to manage wounds in survival situations. Do not try to close any wound by suturing or similar procedures. Leave the wound open to allow the drainage of any pus resulting from infection. As long as the wound can drain, it generally will not become life-threatening, regardless of how unpleasant it looks or smells.

Cover the wound with a clean dressing. Place a bandage on the dressing to hold it in place. Change the dressing daily to check for infection.

If a wound is gaping, you can bring the edges together with adhesive tape cut in the form of a "butterfly" or "dumbbell" (Figure I-7).

In a survival situation, some degree of wound infection is almost inevitable. Pain, swelling, and redness around the wound, increased temperature, and pus in the wound or on the dressing indicate infection is present.

To treat an infected wound—

- Place a warm, moist compress directly on the infected wound.
- Change the compress when it cools, keeping a warm compress on the wound for a total of 30 minutes. Apply the compresses three or four times daily.

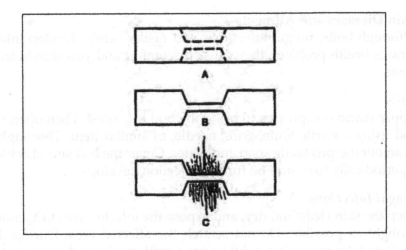

Figure I-7: Butterfly closure

- Drain the wound. Open and gently probe the infected wound with a sterile instrument.
- Dress and bandage the wound.
- Drink a lot of water.

Continue this treatment daily until all signs of infection have disappeared.

If you do not have antibiotics and the wound has become severely infected, does not heal, and ordinary debridement is impossible, consider maggot therapy, despite its hazards:

- Expose the wound to flies for one day and then cover it.
- Check daily for maggots.
- Once maggots develop, keep wound covered but check daily.
- Remove all maggots when they have cleaned out all dead tissue and before they start on healthy tissue. Increased pain and bright red blood in the wound indicate that the maggots have reached healthy tissue.
- Flush the wound repeatedly with sterile water or fresh urine to remove the maggots.
- Check the wound every four hours for several days to ensure all maggots have been removed.
- Bandage the wound and treat it as any other wound. It should heal normally.

Skin Diseases and Ailments

Although boils, fungal infections, and rashes rarely develop into a serious health problem, they cause discomfort and you should treat them.

Boils

Apply warm compresses to bring the boil to a head. Then open the boil using a sterile knife, wire, needle, or similar item. Thoroughly clean out the pus using soap and water. Cover the boil site, checking it periodically to ensure no further infection develops.

Fungal Infections

Keep the skin clean and dry, and expose the infected area to as much sunlight as possible. Do not scratch the affected area. During the Southeast Asian conflict, soldiers used antifungal powders, lye soap, chlorine bleach, alcohol, vinegar, concentrated salt water, and iodine to treat fungal infections with varying degrees of success. As with any "unorthodox" method of treatment, use it with caution.

Rashes

To treat a skin rash effectively, first determine what is causing it. This determination may be difficult even in the best of situations. Observe the following rules to treat rashes:

- If it is moist, keep it dry.
- If it is dry, keep it moist.
- Do not scratch it.

Use a compress of vinegar or tannic acid derived from tea or from boiling acorns or the bark of a hardwood tree to dry weeping rashes. Keep dry rashes moist by rubbing a small amount of rendered animal fat or grease on the affected area.

Remember, treat rashes as open wounds and clean and dress them daily. There are many substances available to survivors in the wild or in captivity for use as antiseptics to treat wound:

- Iodine tablets. Use 5 to 15 tablets in a liter of water to produce a good rinse for wounds during healing.
- Garlic. Rub it on a wound or boil it to extract the oils and use the water to rinse the affected area.
- Salt water. Use 2 to 3 tablespoons per liter of water to kill bacteria.

- Bee honey. Use it straight or dissolved in water.
- Sphagnum moss. Found in boggy areas worldwide, it is a natural source of iodine. Use as a dressing.

Again, use noncommercially prepared materials with caution.

Frostbite
This injury results from frozen tissues. Light frostbite involves only the skin that takes on a dull, whitish pallor. Deep frostbite extends to a depth below the skin. The tissues become solid and immovable. Your feet, hands, and exposed facial areas are particularly vulnerable to frostbite.

When with others, prevent frostbite by using the buddy system. Check your buddy's face often and make sure that he checks yours. If you are alone, periodically cover your nose and lower part of your face with your mittens.

Do not try to thaw the affected areas by placing them close to an open flame. Gently rub them in lukewarm water. Dry the part and place it next to your skin to warm it at body temperature.

Trench Foot
This condition results from many hours or days of exposure to wet or damp conditions at a temperature just above freezing. The nerves and muscles sustain the main damage, but gangrene can occur. In extreme cases the flesh dies and it may become necessary to have the foot or leg amputated. The best prevention is to keep your feet dry. Carry extra socks with you in a waterproof packet. Dry wet socks against your body. Wash your feet daily and put on dry socks.

Burns
The following field treatment for burns relieves the pain somewhat, seems to help speed healing, and offers some protection against infection:

- First, stop the burning process. Put out the fire by removing clothing, dousing with water or sand, or by rolling on the ground. Cool the burning skin with ice or water. For burns caused by white phosphorous, pick out the white phosphorous with tweezers; do not douse with water.
- Soak dressings or clean rags for 10 minutes in a boiling tannic acid solution (obtained from tea, inner bark of hardwood trees, or acorns boiled in water).

- Cool the dressings or clean rags and apply over burns.
- Rest as an open wound.
- Replace fluid loss.
- Maintain airway.
- Treat for shock.
- Consider using morphine, unless the burns are near the face.

ENVIRONMENTAL INJURIES

Heatstroke, hypothermia, diarrhea, and intestinal parasites are environmental injuries you could face.

Heatstroke

The breakdown of the body's heat regulatory system (body temperature more than 40.5 degrees C [105 degrees F]) causes a heatstroke. Other heat injuries, such as cramps or dehydration, do not always precede a heatstroke. Signs and symptoms of heatstroke are—

- Swollen, beet-red face.
- Reddened whites of eyes.
- Victim not sweating.
- Unconsciousness or delirium, which can cause pallor, a bluish color to lips and nail beds (cyanosis), and cool skin.

Note: By this time the victim is in severe shock. Cool the victim as rapidly as possible. Cool him by dipping him in a cool stream. If one is not available, douse the victim with urine, water, or at the very least, apply cool wet compresses to all the joints, especially the neck, armpits, and crotch. Be sure to wet the victim's head. Heat loss through the scalp is great. Administer IVs and provide drinking fluids. You may fan the individual.

Expect, during cooling—

- Vomiting.
- Diarrhea.
- Struggling.
- Shivering.
- Shouting.
- Prolonged unconsciousness.
- Rebound heatstroke within 48 hours.
- Cardiac arrest; be ready to perform CPR.
- Prolonged unconsciousness.

- Rebound heatstroke within 48 hours.
- Cardiac arrest; be ready to perform CPR.

Note: Treat for dehydration with lightly salted water.

Hypothermia

Defined as the body's failure to maintain a temperature of 36 degrees C (97 degrees F). Exposure to cool or cold temperature over a short or long time can cause hypothermia. Dehydration and lack of food and rest predispose the survivor to hypothermia.

Unlike heatstroke, you must gradually warm the hypothermia victim. Get the victim into dry clothing. Replace lost fluids, and warm him.

Diarrhea

A common, debilitating ailment caused by a change of water and food, drinking contaminated water, eating spoiled food, becoming fatigued, and using dirty dishes. You can avoid most of these causes by practicing preventive medicine. If you get diarrhea, however, and do not have anti-diarrheal medicine, one of the following treatments may be effective:

- Limit your intake of fluids for 24 hours.
- Drink one cup of a strong tea solution every 2 hours until the diarrhea slows or stops. The tannic acid in the tea helps to control the diarrhea. Boil the inner bark of a hardwood tree for 2 hours or more to release the tannic acid.
- Make a solution of one handful of ground chalk, charcoal, or dried bones and treated water. If you have some apple pomace or the rind of citrus fruit, add an equal portion to the mixture to make it more effective. Take 2 tablespoons of the solution every 2 hours until the diarrhea slows or stops.

Intestinal Parasites

You can usually avoid worm infestations and other intestinal parasites if you take preventive measures. For example, never go barefoot. The most effective way to prevent intestinal parasites is to avoid uncooked meat and raw vegetables contaminated by raw sewage or human waste used as a fertilizer. However, should you become infested and lack proper medicine, you can use home remedies. Keep in mind that these home remedies work on the principle of changing the environment of the gastrointestinal tract. The following are home remedies you could use:

- Salt water. Dissolve 4 tablespoons of salt in 1 liter of water and drink. Do not repeat this treatment.
- Tobacco. Eat 1 to 1.5 cigarettes. The nicotine in the cigarette will kill or stun the worms long enough for your system to pass them. If the infestation is severe, repeat the treatment in 24 to 48 hours, but no sooner.
- Kerosene. Drink 2 tablespoons of kerosene but no more. If necessary, you can repeat this treatment in 24 to 48 hours. Be careful not to inhale the fumes. They may cause lung irritation.
- Hot peppers. Peppers are effective only if they are a steady part of your diet. You can eat them raw or put them in soups or rice and meat dishes. They create an environment that is prohibitive to parasitic attachment.

HERBAL MEDICINES

Our modern wonder drugs, laboratories, and equipment have obscured more primitive types of medicine involving determination, commonsense, and a few simple treatments. In many areas of the world, however, the people still depend on local "witch doctors" or healers to curetheir ailments. Many of the herbs (plants) and treatments they use areas effective as the most modern medications available. In fact, many modern medications come from refined herbs.

> **WARNING**
> Use herbal medicines with extreme care, however, and only when you lack or have limited medical supplies. Some herbal medicines are dangerous and may cause further damage or even death.

CHAPTER 1

FUNDAMENTAL CRITERIA FOR FIRST AID

Soldiers may have to depend upon their first aid knowledge and skills to save themselves or other soldiers. They may be able to save a life, prevent permanent disability, and reduce long periods of hospitalization by knowing what to do, what not to do, and when to seek medical assistance. Anything soldiers can do to keep others in good fighting condition is part of the primary mission to fight or to support the weapons system. Most injured or ill soldiers are able to return to their units to fight and/or support primarily because they are given appropriate and timely first aid followed by the best medical care possible. Therefore, all soldiers must remember the basics:

- Check for BREATHING: Lack of oxygen intake (through a compromised airway or inadequate breathing) can lead to brain damage or death in very few minutes.
- Check for BLEEDING: Life cannot continue without an adequate volume of blood to carry oxygen to tissues.
- Check for SHOCK: Unless shock is prevented or treated, death may result even though the injury would not otherwise be fatal.

SECTION I. EVALUATE CASUALTY

1-1. Casualty Evaluation (081-831-1000)

The time may come when you must instantly apply your knowledge of lifesaving and first aid measures, possibly under combat or other adverse conditions. Any soldier observing an unconscious and/or ill, injured, or wounded person must carefully and skillfully evaluate him to determine the first aid measures required to prevent further injury or death. He should seek help from medical personnel as soon as possible, but must NOT interrupt his evaluation or treatment of the casualty. A second person may be sent to find medical help. One of the cardinal principles of treating a casualty is that the initial rescuer must continue the evaluation and treatment, as the tactical situation permits, until he is relieved by another individual. If, during any part of the evaluation, the casualty exhibits the conditions for which the soldier is checking, the soldier must stop the evaluation and immediately administer first aid. Ina chemical environment, the soldier should not evaluate the casualty until the casualty has been masked and given the antidote. After providing first aid, the soldier must proceed with the evaluation and continue to monitor the casualty for further medical complications until relieved by medical personnel. Learn the following procedures well. You may become that soldier who will have to give first aid some day.

✍ NOTE

Remember, when evaluating and/or treating a casualty, you should seek medical aid as soon as possible. DO NOT stop treatment, but if the situation allows, send another person to find medical aid.

WARNING

Again, remember, if there are any signs of chemical or biological agent poisoning, you should immediately mask the casualty. If it is nerve agent poisoning, administer the antidote, using the casualty's injector/ampules. See task081-831-1031, Administer First Aid to a Nerve Agent Casualty (Buddy Aid).

---

a. Step ONE. Check the casualty for responsiveness by gently shaking or tapping him while calmly asking, "Are you okay?" Watch for response. If the casualty does not respond, go to step TWO. See Chapter2, paragraph 2-5 for more information. If the casualty responds, continue with the evaluation.

(1) If the casualty is conscious, ask him where he feels different than usual or where it hurts. Ask him to identify

the locations of pain if he can, or to identify the area in which there is no feeling.

(2) If the casualty is conscious but is choking and cannot talk, stop the evaluation and begin treatment. See task 081-831-1003Clear an Object from the Throat of a Conscious Casualty. Also see Chapter 2, paragraph 2-13 for specific details on opening the airway.

WARNING

If a broken neck or back is suspected, do not move the casualty unless to save his life. Movement may cause permanent paralysis or death.

b. Step TWO. Check for breathing. See Chapter 2, paragraph 2-5c for procedure.

(1) If the casualty is breathing, proceed to step FOUR.

(2) If the casualty is not breathing, stop the evaluation and begin treatment (attempt to ventilate). See task 081-831-1042, Perform Mouth-to-Mouth Resuscitation. If an airway obstruction is apparent, clear the airway obstruction, then ventilate.

(3) After successfully clearing the casualty's airway, proceed to step THREE.

c. Step THREE. Check for pulse. If pulse is present, and the casualty is breathing, proceed to step FOUR.

(1) If pulse is present, but the casualty is still not breathing, start rescue breathing. See Chapter 2, paragraphs 2-6, and 2-7 for specific methods.

*(2) If pulse is not found, seek medically trained personnel for help.

d. Step FOUR. Check for bleeding. Look for spurts of blood or blood-soaked clothes. Also check for both entry and exit wounds. If the casualty is bleeding from an open wound, stop the evaluation and begin first aid treatment in accordance with the following tasks, as appropriate:

(1) Arm or leg wound–Task 081-831-1016, Put on a Field or Pressure Dressing. See Chapter 2, paragraphs 2-15, 2-17, 2-18, and 2-19.

(2) Partial or complete amputation–Task 081-831-1017,Put on a Tourniquet. See Chapter 2, paragraph 2-20.

(3) Open head wound–Task 081-831-1033, Apply a Dressing to an Open Head Wound. See Chapter 3, Section I.

(4) Open abdominal wound–Task 081-831-1025,Apply a Dressing to an Open Abdominal Wound. See Chapter 3, paragraph 3-12.

(5) Open chest wound–Task 081-831-1026, Apply a Dressing to an Open Chest Wound. See Chapter 3, paragraphs 3-9 and 3-10.

> **WARNING**
> In a chemically contaminated area, do not expose the wound(s).

e. Step FIVE. Check for shock. If signs/symptoms of shock are present, stop the evaluation and begin treatment immediately. The following are nine signs and/or symptoms of shock.

(1) Sweaty but cool skin (clammy skin).

(2) Paleness of skin.

(3) Restlessness or nervousness.

(4) Thirst.

(5) Loss of blood (bleeding).

(6) Confusion (does not seem aware of surroundings).

(7) Faster than normal breathing rate.

(8) Blotchy or bluish skin, especially around the mouth.

(9) Nausea and/or vomiting.

> **WARNING**
> Leg fractures must be splinted before elevating the legs/as a treatment for shock.

See Chapter 2, Section III for specific information regarding the causes and effects, signs/symptoms, and the treatment/prevention of shock.

---

   f. Step SIX. Check for fractures (Chapter 4).
   (1) Check for the following signs/symptoms of a back or neck injury and treat as necessary.
   - Pain or tenderness of the neck or back area.
   - Cuts or bruises in the neck or back area.
   - Inability of a casualty to move (paralysis or numbness).
     o Ask about ability to move (paralysis).
     o Touch the casualty's arms and legs and ask whether he can feel your hand (numbness).
   - Unusual body or limb position.

WARNING

Unless there is immediate life threatening danger, do not move a casualty who has a suspected back or neck injury. Movement may cause permanent paralysis or death.

   (2) Immobilize any casualty suspected of having a neck or back injury by doing the following
   - Tell the casualty not to move.
   - If a back injury is suspected, place padding (rolled or folded to conform to the shape of the arch) under the natural arch of the casualty's back. For example, a blanket may be used as padding.
   - If a neck injury is suspected, place a roll of cloth under the casualty's neck and put weighted boots (filled with dirt, sand and so forth) or rocks on both sides of his head.
   (3) Check the casualty's arms and legs for open or closed fractures.
   Check for open fractures.
   Look for bleeding.
   Look for bone sticking through the skin.

> Check for closed fractures.
> Look for swelling.
> Look for discoloration.
> Look for deformity.
> Look for unusual body position.
> *(4) Stop the evaluation and begin treatment if a fracture to an arm or leg is suspected. See Task 081-831-1034, Splint a Suspected Fracture, Chapter 4, paragraphs 4-4 through 4-7.

(5) Check for signs/symptoms of fractures of other body areas (for example, shoulder or hip) and treat as necessary.

> g. Step SEVEN. Check for burns. Look carefully for reddened, blistered, or charred skin, also check for singed clothing. If bums are found, stop the evaluation and begin treatment (Chapter 3, paragraph 3-14). See task 081-831-1007, *Give First Aid for Burns*.
> h. Step EIGHT. Check for possible head injury.
>   (1) Look for the following signs and symptoms
>   Unequal pupils.
>   Fluid from the ear(s), nose, mouth, or injury site.
>   Slurred speech.
>   Confusion.
>   Sleepiness.
>   Loss of memory or consciousness.
>   Staggering in walking.
>   Headache.
>   Dizziness.
>   Vomiting and/or nausea.
>   Paralysis.
>   Convulsions or twitches.

(2) If a head injury is suspected, continue to watch for signs which would require performance of mouth-to-mouth resuscitation, treatment for shock, or control of bleeding and seek medical aid. See Chapter 3, Section I for specific indications of head injury and treatment. See task 081-831-1033, Apply a Dressing to an Open Head Wound.

1-2. Medical Assistance (081-831-1000)

When a non-medically trained soldier comes upon an unconscious and/or injured soldier, he/she must accurately evaluate the casualty to determine the first aid measures needed to prevent further injury or death. He/she should seek medical assistance as soon as possible, but he/she MUST NOT interrupt treatment. To interrupt treatment may cause more harm than good to the casualty. A second person may be sent to find medical help. If, during anypart of the evaluation, the casualty exhibits the conditions for which the soldier is checking, the soldier must stop the evaluation and immediately administer first aid. Remember that in a chemical environment, the soldier should not evaluate the casualty until the casualty has been masked and given the antidote. After performing first aid, the soldier must proceed with the evaluation and continue to monitor the casualty for development of conditions which may require the performance of necessary basic life saving measures, such as clearing the airway, mouth-to-mouth resuscitation, preventing shock, ardor bleeding control. He/she should continue to monitor until relieved by medical personnel.

SECTION II. UNDERSTAND VITAL BODY FUNCTIONS

1-3. Respiration and Blood Circulation

Respiration (inhalation and exhalation) and blood circulation are vital body functions. Interruption of either of these two functions need not be fatal IF appropriate first aid measures are correctly applied.

a. Respiration. When a person inhales, oxygen is taken into the body and when he exhales, carbon dioxide is expelled from the body–this is respiration. Respiration involves the—

- Airway (nose, mouth, throat, voice box, windpipe, and bronchial tree). The canal through which air passes to and from the lungs.
- Lungs (two elastic organs made up of thousands of tiny air spaces and covered by an airtight membrane).
- Chest cage (formed by the muscle-connected ribs which join the spine in back and the breastbone in front). The top part of the chest cage is closed by the structure of the neck, and the bottom part is separated from the abdominal cavity by a large dome-shaped muscle called the diaphragm (Figure 1-1). The diaphragm and rib muscles, which are under the control of the respiratory center in the brain, automatically contract and relax. Contraction increases and relaxation decreases the size of the chest cage.

When the chest cage increases and then decreases, the air pressure in the lungs is first less and then more than the atmospheric pressure, thus causing the air to rush in and out of the lungs to equalize the pressure. This cycle of inhaling and exhaling is repeated about 12 to 18 times per minute.

b. Blood Circulation. The heart and the blood vessels (arteries, veins, and capillaries) circulate blood through the body tissues. The heart is divided into two separate halves, each acting as a pump. The left side pumps oxygenated blood (bright red) through the arteries into the capillaries; nutrients and oxygen pass from the blood through the walls of the capillaries into the cells. At the same time waste products and carbon dioxide enter the capillaries. From the capillaries the oxygen poor blood is carried through the veins to the right side of the heart and then into the lungs where it expels carbon dioxide and picks up oxygen, Blood in the veins is dark red because of its low oxygen content. Blood does not flow through the veins in spurts as it does through the arteries.

(1) Heartbeat. The heart functions as a pump to circulate the blood continuously through the blood vessels to all parts of the body. It contracts, forcing the blood from its chambers; then it relaxes, permitting its chambers to refill with blood. The rhythmical cycle of contraction and relaxation

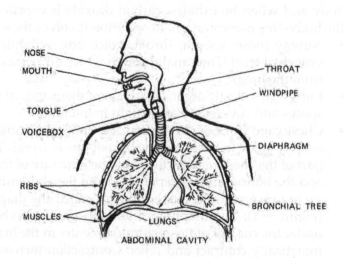

Figure 1-1: Airway, lungs, and chest cage

is called the heartbeat. The normal heartbeat is from 60 to 80 beats per minute.

(2) Pulse. The heartbeat causes a rhythmical expansion and contraction of the arteries as it forces blood through them. This cycle of expansion and contraction can be felt (monitored) at various body points and is called the pulse. The common points for checking the pulse are at the side of the neck (carotid), the groin (femoral), the wrist (radial), and the ankle (posterial tibial).

(a) Neck (carotid) pulse. To check the neck (carotid) pulse, feel for a pulse on the side of the casualty's neck closest to you by placing the tips of your first two fingers beside his Adam's apple (Figure 1-2).

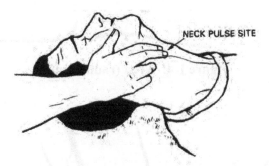

Figure 1-2: Neck (carotid) pulse

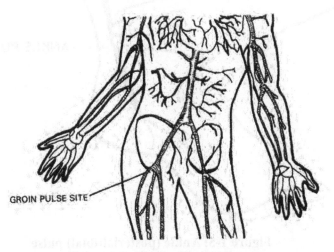

Figure 1-3: Groin (femoral) pulse

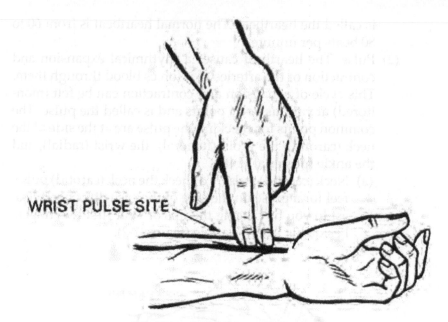

WRIST PULSE SITE

Figure 1-4: Wrist (radial) pulse

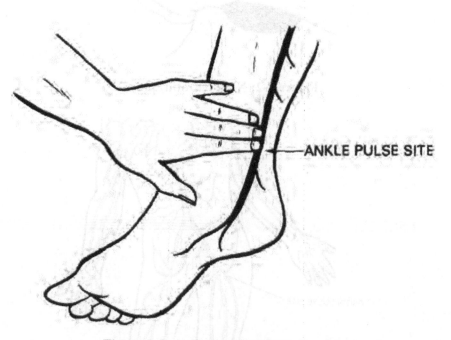

ANKLE PULSE SITE

Figure 1-5: Ankle (posterialtibial) pulse

(b) Groin (femoral) pulse. To check the groin (femoral) pulse, press the tips of two fingers into the middle of the groin (Figure 1-3).

(c) Wrist (radial) pulse. To check the wrist (radial) pulse, place your first two fingers on the thumb side of the casualty's wrist (Figure 1-4).

(d) Ankle (posterial tibial) pulse. To check the ankle (posterial tibial) pulse, place your first two fingers on the inside of the ankle (Figure 1-5).

✍ NOTE

DO NOT use your thumb to check a casualty's pulse because you may confuse your pulse beat with that of the casualty.

1-4. Adverse Conditions

a. *Lack of Oxygen*. Human life cannot exist without a continuous intake of oxygen. Lack of oxygen rapidly leads to death. First aid involves knowing how to OPEN THE AIRWAY AND RE-STORE BREATHING AND HEARTBEAT (Chapter 2, Section I).

b. *Bleeding*. Human life cannot continue without an adequate volume of blood to carry oxygen to the tissues. An important first aid measure is to STOP THE BLEEDING to prevent loss of blood (Chapter 2, Section II).

c. *Shock*. Shock means there is inadequate blood flow to the vital tissues and organs. Shock that remains uncorrected may result in death even though the injury or condition causing the shock would not otherwise be fatal. Shock can result from many causes, such as loss of blood, loss of fluid from deep burns, pain, and reaction to the sight of a wound or blood. First aid includes PREVENTING SHOCK, since the casualty's chances of survival are much greater if he does not develop shock (Chapter 2, Section III).

d. *Infection*. Recovery from a severe injury or a wound depends largely upon how well the injury or wound was initially protected. Infections result from the multiplication and growth (spread) of germs (bacteria: harmful microscopic organisms). Since harmful bacteria are in the air and on the skin and clothing, some of these organisms will immediately invade (contaminate) a break in the skin or an open wound. The objective is to KEEP ADDITIONAL GERMS OUT OF THE WOUND. A good working knowledge of basic first aid measures also includes knowing how to dress the wound to avoid infection or additional contamination (Chapters 2 and 3).

CHAPTER 2

BASIC MEASURES FOR FIRST AID

Several conditions which require immediate attention are an inadequate airway, lack of breathing or lack of heartbeat, and excessive loss of blood. A casualty without a clear airway or who is not breathing may die from lack of oxygen. Excessive loss of blood may lead to shock, and shock can lead to death; therefore, you must act immediately to control the loss of blood. All wounds are considered to be contaminated, since infection-producing organisms (germs) are always present on the skin, on clothing, and in the air. Any missile or instrument causing the wound pushes or carries the germs into the wound. Infection results as these organisms multiply. That a wound is contaminated does not lessen the importance of protecting it from further contamination. You must dress and bandage a wound as soon as possible to prevent further contamination. It is also important that you attend to any airway, breathing, or bleeding problem IMMEDIATELY because these problems may become life-threatening.

SECTION I. OPEN THE AIRWAY AND RESTORE BREATHING

*2-1. Breathing Process
All living things must have oxygen to live. Through the breathing process, the lungs draw oxygen from the air and put it into the blood. The heart pumps the blood through the body to be used by the living cells which require a constant supply of oxygen. Some cells are more dependent on a constant supply of oxygen than others. Cells of the brain may die within 4 to 6 minutes without oxygen. Once these cells die, they are lost forever since they DO NOT regenerate. This could result in permanent brain damage, paralysis, or death.

41

SOURCE: Copyright. American Heart Association. *Instructor's Manual for Basic Life Support.* Dallas: American Heart Association, 1987.

Figure 2-1: Responsiveness checked

2-2. Assessment (Evaluation) Phase (081-831-1000 and 081-831-1042)

  a. Check for responsiveness (Figure 2-1A)—establish whether the casualty is conscious by gently shaking him and asking, "Are you O.K.?"
  b. Call for help (Figure 2-1B).
  c. Position the unconscious casualty so that he is lying on his back and on a firm surface (Figure 2-1C) (081-831-1042).

WARNING (081-831-1042)
If the casualty is lying on his chest (prone position), cautiously roll the casualty as a unit so that his body does not twist(which may further complicate a neck, back or spinal injury).

(1) Straighten the casualty's legs. Take the casualty's arm that is nearest to you and move it so that it is straight and above his head. Repeat procedure for the other arm.

(2) Kneel beside the casualty with your knees near his shoulders (leave space to roll his body) (Figure 2-1B). Place one hand behind his head and neck for support. With your other hand, grasp the casualty under his far arm (Figure 2-1C).

(3) Roll the casualty toward you using a steady and even pull. His head and neck should stay in line with his back.

(4) Return the casualty's arms to his sides. Straighten his legs. Reposition yourself so that you are now kneeling at the level of the casualty's shoulders. However, if a neck injury is suspected, and the jaw-thrust will be used, kneel at the casualty's head, looking toward his feet.

2-3. Opening the Airway—Unconscious and Not Breathing Casualty (081-831-1042)

*The tongue is the single most common cause of an airway obstruction (Figure 2-2). In most cases, the airway can be cleared by simply using the head-tilt/chin-lift technique. This action pulls the tongue away from the air passage in the throat (Figure 2-3).

a. Step ONE (081-331-1042). Call for help and then position the casualty. Move (roll) the casualty onto his back (Figure 2-1C above).

> ⚠ **CAUTION (081-831-1007)**
> Take care in moving a casualty with a suspected neck or back injury. Moving an injured neck or back may permanently injure the spine.

SOURCE: Copyright. American Heart Association. *Instructor's Manual for Basic Life Support.* Dallas: American Heart Association, 1987.

Figure 2-2: Airway blocked by tongue

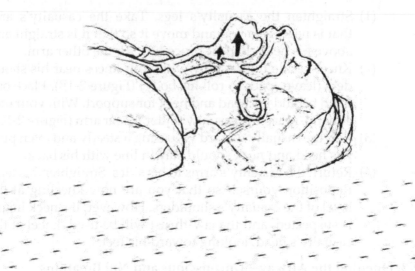

Figure 2-3: Airway opened (cleared)

✍ NOTE (081-831-1042)

If foreign material or vomit is visible in the mouth, it should be removed, but do not spend an excessive amount of time doing so.

---

b. Step TWO (081-831-1042). Open the airway using the jaw-thrust or head-tilt/chin-lift technique.

✍ NOTE

The head-tilt/chin-lift is an important procedure in opening the airway; however, use extreme care because excess force in performing this maneuver may cause further spinal injury. In a casualty with a suspected neck injury or severe head trauma, the safest approach to opening the airway is the jaw-thrust technique because in most cases it can be accomplished without extending the neck.

---

(1) Perform the jaw-thrust technique. The jaw-thrust maybe accomplished by the rescuer grasping the angles of the casualty's lower jaw and lifting with both hands, one on

Figure 2-4. Jaw-thrust technique of opening airway

each side, displacing the jaw forward and up (Figure 2-4).
The rescuer's elbows should rest on the surface on which
the casualty is lying. If the lips close, the lower lip can be
retracted with the thumb. If mouth-to-mouth breathing is
necessary, close the nostrils by placing your cheek tightly against them. The head should be carefully supported
without tilting it backwards or turning it from side to side.
If this is unsuccessful, the head should be tilted back very
slightly. The jaw-thrust is the safest first approach to opening the airway of a casualty who has a suspected neck injury because in most cases it can be accomplished without
extending the neck.

(2) Perform the head-tilt/chin-lift technique (081-831-1042).
Place one hand on the casualty's forehead and apply firm,
backward pressure with the palm to tilt the head back.
Place the fingertips of the other hand under the bony part
of the lower jaw and lift, bringing the chin forward. The
thumb should not be used to lift the chin (Figure 2-5).

✍ NOTE

The fingers should not press deeply into the soft tissue
under the chin because the airway may be obstructed.

---

c. Step THREE. Check for breathing (while maintaining an airway). After establishing an open airway, it is important to
maintain that airway in an open position. Often the act of just
opening and maintaining the airway will allow the casualty to
breathe properly. Once the rescuer uses one of the techniques

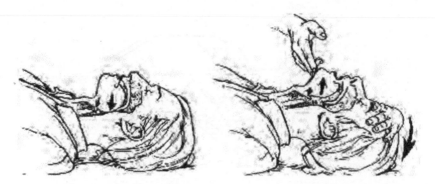

Figure 2-5. Head-tilt/chin-lift technique of opening airway

to open the airway (jaw-thrust or head-tilt/chin-lift), he should maintain that head position to keep the airway open. Failure to maintain the open airway will prevent the casualty from receiving an adequate supply of oxygen. Therefore, while maintaining an open airway, the rescuer should check for breathing by observing the casualty's chest and performing the following actions within 3 to 5 seconds:

(1) LOOK for the chest to rise and fall.
(2) LISTEN for air escaping during exhalation by placing your ear near the casualty's mouth.
(3) FEEL for the flow of air on your cheek (see Figure 2-6),
(4) If the casualty does not resume breathing, give mouth-to-mouth resuscitation.

✍ NOTE

If the casualty resumes breathing, monitor and maintain the open airway. If he continues to breathe, he should be transported to a medical treatment facility.

2-4. Rescue Breathing (Artificial Respiration)

a. If the casualty does not promptly resume adequate spontaneous breathing after the airway is open, rescue breathing(artificial respiration) must be started. Be calm! Think and act quickly! The sooner you begin rescue breathing, the more likely you are to restore the casualty's breathing. If you are in doubt whether the casualty is breathing, give artificial respiration, since it can do no harm to a person who is breathing. If the casualty is breathing, you can feel and see his chest move. Also, if the casualty is breathing, you can feel and hear air being expelled by putting your hand or ear close to his mouth and nose.

b. There are several methods of administering rescue breathing. The mouth-to-mouth method is preferred; however, it cannot be used in all situations. If the casualty has a severe jaw fracture or mouth wound or his jaws are tightly closed by spasms, use the mouth-to-nose method.

2-5. Preliminary Steps—All Rescue Breathing Methods (081-831-1042)

a. Step ONE. Establish unresponsiveness. Call for help. Turn or position the casualty.

b. Step TWO. Open the airway.

c. Step THREE. Check for breathing by placing your ear over the casualty's mouth and nose, and looking toward his chest:

(1) **Look** for rise and fall of the casualty's chest (Figure 2-6).

(2) **Listen** for sounds of breathing.

(3) **Feel** for breath on the side of your face. If the chest does not rise and fall and no air is exhaled, then the casualty is breathless (not breathing). (This evaluation procedure should take only 3 to 5 seconds. Perform rescue breathing if the casualty is not breathing.

Figure 2-6

> ✍ **NOTE**
> Although the rescuer may notice that the casualty is making respiratory efforts, the airway may still be obstructed and opening the airway may be all that is needed. If the casualty resumes breathing, the rescuer should continue to help maintain an open airway.

2-6. Mouth-to-Mouth Method (081-831-1042)

In this method of rescue breathing, you inflate the casualty's lungs with air from your lungs. This can be accomplished by blowing air into the person's mouth. The mouth-to-mouth rescue breathing method is performed as follows:

a. Preliminary Steps.

   (1) Step ONE (081-831-1042). If the casualty is not breathing, place your hand on his forehead, and pinch his nostrils together with the thumb and index finger of this same hand. Let this same hand exert pressure on his forehead to maintain the backward head-tilt and maintain an open airway. With your other hand, keep your fingertips on the bony part of the lower jaw near the chin and lift (Figure 2-7).

> ✍ **NOTE**
> If you suspect the casualty has a neck injury and you are using the jaw-thrust technique, close the nostrils by placing your cheek tightly against them.

Figure 2-7: Head-tilt/chin-lift

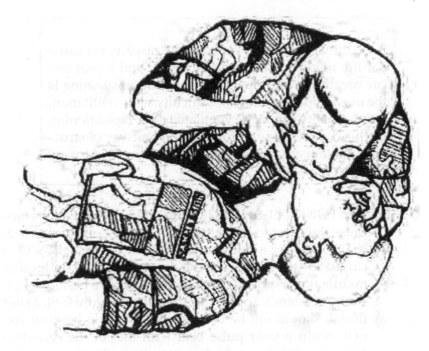

Figure 2-8: Rescue breathing

(2) Step TWO (081-831-1042).Take a deep breath and place your mouth (in an airtight seal) around the casualty's mouth (Figure 2-8). (If the injured person is small, cover both his nose and mouth with your mouth, sealing your lips against the skin of his face.)

(3) Step THREE (081-831-1042). Blow two full breaths into the casualty's mouth (1 to 1 1/2 seconds per breath), taking a breath of fresh air each time before you blow. Watch out of the corner of your eye for the casualty's chest to rise. If the chest rises, sufficient air is getting into the casualty's lungs. Therefore, proceed as described in step FOUR below. If the chest does not rise, do the following (a, b, and c below) and then attempt to ventilate again.

(a) Take corrective action immediately by reestablishing the airway. Make sure that air is not leaking from around your mouth or out of the casualty's pinched nose.

(b) Reattempt to ventilate.

(c) If chest still does not rise, take the necessary action to open an obstructed airway (paragraph 2-14).

✍ NOTE

If the initial attempt to ventilate the casualty is unsuccessful, reposition the casualty's head and repeat rescue breathing. Improper chin and head positioning is
the most, common cause of difficulty with ventilation.
If the casualty cannot be ventilated after repositioning
the head, proceed with foreign-body airway obstruction maneuvers (see Open an Obstructed Airway,
paragraph 2-14).

---

(4) Step FOUR (081-831-1042). After giving two breaths which
cause the chest to rise, attempt to locate a pulse on the casualty. Feel for a pulse on the side of the casualty's neck
closest to you by placing the first two fingers (index and
middle fingers) of your hand on the groove beside the casualty's Adam's apple (carotid pulse) (Figure 2-9). (Your
thumb should not be used for pulse taking because you
may confuse your pulse beat with that of the casualty.)
Maintain the airway by keeping your other hand on the
casualty's forehead. Allow 5 to 10 seconds to determine if
there is a pulse.

(a) If a pulse is found and the casualty is breathing—STOP
allow the casualty to breathe on his own. If possible, keep
him warm and comfortable.

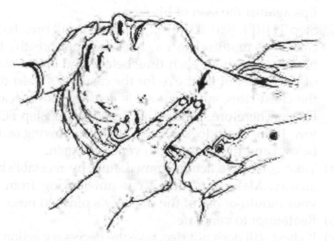

Figure 2-9. Placement of fingers to detect pulse

    (b) If a pulse is found and the casualty is not breathing, continue rescue breathing.

   * (c) If a pulse is not found, seek medically trained personnel for help.

  b. Rescue Breathing (mouth-to-mouth resuscitation) (081-831-1042). Rescue breathing (mouth-to-mouth or mouth-to-nose resuscitation) is performed at the rate of about one breath every 5 seconds (12 breaths per minute) with rechecks for pulse and breathing after every 12 breaths. Rechecks can be accomplished in 3 to 5 seconds. See steps ONE through SEVEN (below) for specifics.

> ✎ **NOTE**
> Seek help (medical aid), if not done previously.

   (1) Step ONE. If the casualty is not breathing, pinch his nostrils together with the thumb and index finger of the hand on his forehead and let this same hand exert pressure on the forehead to maintain the backward head-tilt (Figure 2-7).

   (2) Step TWO. Take a deep breath and place your mouth(in an airtight seal) around the casualty's mouth (Figure 2-8).

   (3) Step THREE. Blow a quick breath into the casualty's mouth forcefully to cause his chest to rise. If the casualty's chest rises, sufficient air is getting into his lungs.

   (4) Step FOUR. When the casualty's chest rises, remove your mouth from his mouth and listen for the return of air from his lungs(exhalation).

   (5) Step FIVE. Repeat this procedure (mouth-to-mouth resuscitation) at a rate of one breath every 5 seconds to achieve 12 breaths per minute. Use the following count: "one, one-thousand; two, one-thousand; three, one-thousand; four, one-thousand; BREATH; one, one-thousand;" and so forth. To achieve a rate of one breath every 5seconds, the breath must be given on the fifth count.

  * (6) Step SIX. Feel for a pulse after every twelfth breath. This check should take about 3 to 5 seconds. If a pulse beat is not found, seek medically trained personnel for help.

  * (7) Step SEVEN. Continue rescue breathing until the casualty starts to breathe on his own, until you are relieved by

another person, or until you are too tired to continue. Monitor pulse and return of spontaneous breathing after every few minutes of rescue breathing. If spontaneous breathing returns, monitor the casualty closely. The casualty should then be transported to a medical treatment facility. Maintain an open airway and be prepared to resume rescue breathing, if necessary.

2-7. Mouth-to-Nose Method

Use this method if you cannot perform mouth-to-mouth rescue breathing because the casualty has a severe jaw fracture or mouth wound or his jaws are tightly closed by spasms. The mouth-to-nose method is performed in the same way as the mouth-to-mouth method except that you blow into the nose while you hold the lips closed with one hand at the chin. You then remove your mouth to allow the casualty to exhale passively. It may be necessary to separate the casualty's lips to allow the air to escape during exhalation.

* 2-8. Heartbeat

If a casualty's heart stops beating, you must immediately seek medically trained personnel for help. SECONDS COUNT! Stoppage of the heart is soon followed by cessation of respiration unless it has occurred first. Becalm! Think and act! When a casualty's heart has stopped, there is no pulse at all; the person is unconscious and limp, and the pupils of his eyes are open wide. When evaluating a casualty or when performing the preliminary steps of rescue breathing, feel for a pulse. If you DO NOT detect a pulse, immediately seek medically trained personnel.

Note: The U.S. Army deleted paragraphs 2-9, 2-10, and 2-11 of this manual as part of a revision.

2-12. Airway Obstructions

In order for oxygen from the air to flow to and from the lungs, the upper airway must be unobstructed.

    a. Upper airway obstructions often occur because—
        (1) The casualty's tongue falls back into his throat while he is unconscious as a result of injury, cardiopulmonary arrest, and so forth. (The tongue falls back and obstructs, it is not swallowed.)
        (2) Foreign bodies become lodged in the throat. These obstructions usually occur while eating (meat most

commonly causes obstructions). Choking on food is associated with—

- Attempting to swallow large pieces of poorly chewed food.
- Drinking alcohol.
- Slipping dentures.

(3) The contents of the stomach are regurgitated and may block the airway.

(4) Blood clots may form as a result of head and facial injuries.

b. Upper airway obstructions may be prevented by taking the following precautions:

(1) Cut food into small pieces and take care to chew slowly and thoroughly.

(2) Avoid laughing and talking when chewing and swallowing.

(3) Restrict alcohol while eating meals.

(4) Keep food and foreign objects from children while they walk, run, or play.

(5) Consider the correct positioning/maintenance of the open airway for the injured or unconscious casualty.

c. Upper airway obstruction may cause either *partial* or *complete* airway blockage.

* (1) *Partial airway obstruction.* The casualty may still have an air exchange. A good air exchange means that the casualty can cough forcefully, though he may be wheezing between coughs. You, the rescuer, should not interfere, and should encourage the casualty to cough up the object on his own. A poor air exchange may be indicated by weak coughing with a high pitched noise between coughs. Additionally, the casualty may show signs of shock (for example, paleness of the skin, bluish or grayish tint around the lips or fingernail beds) indicating a need for oxygen. You should assist the casualty and treat him as though he had a complete obstruction.

(2) *Complete airway obstruction.* A complete obstruction (no air exchange) is indicated if the casualty cannot speak, breathe, or cough at all. He may be clutching his neck and moving erratically. In an unconscious casualty a complete obstruction is also indicated if after opening his airway you cannot ventilate him.

2-13. Opening the Obstructed Airway-Conscious Casualty (081-831-1003)

Clearing a conscious casualty's airway obstruction can be performed with the casualty either standing or sitting, and by following a relatively simple procedure.

> **WARNING**
>
> Once an obstructed airway occurs, the brain will develop an oxygen deficiency resulting in unconsciousness. Death will follow rapidly if prompt action is not taken.

    a. Step ONE. Ask the casualty if he can speak or if he is choking. Check for the universal choking sign (Figure 2-18).

    b. Step TWO. If the casualty can speak, encourage him to attempt to cough; the casualty still has a good air exchange. If he is able to speak or cough effectively, DO NOT interfere with his attempts to expel the obstruction.

    c. Step THREE. Listen for high pitched sounds when the casualty breathes or coughs (poor air exchange). If there is poor air exchange or no breathing, CALL for HELP and immediately deliver manual thrusts (either an abdominal or chest thrust).

Figure 2-18: Universal sign of choking

✍ **NOTE**

The manual thrust with the hands centered between the waist, and the rib cage is called an abdominal thrust (or Heimlich maneuver). The chest thrust (the hands are centered in the middle of the breastbone) is used only for an individual in the advanced stages of pregnancy, in the markedly obese casualty, or if there is a significant abdominal wound.

- Apply ABDOMINAL THRUSTS using the procedures below:
  - ○ Stand behind the casualty and wrap your arms around his waist. Make a fist with one hand and grasp it with the other. The thumb side of your fist should be against the casualty's abdomen, in the midline and slightly above the casualty's navel, but well below the tip of the breastbone (Figure 2-19).
  - ○ Press the fists into the abdomen with a quick backward and upward thrust (Figure 2-20).
  - ○ Each thrust should be a separate and distinct movement.

✍ ★ **NOTE**

Continue performing abdominal thrusts until the obstruction is expelled or the casualty becomes unconscious.

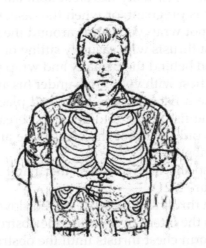

Figure 2-19: Anatomical view of abdominal thrust procedure

Figure 2-20: Profile view of abdominal thrust

- o If the casualty becomes unconscious, call for help as you proceed with steps to open the airway and perform rescue breathing (See task 081-831-1042, Perform Mouth-to-Mouth Resuscitation.)
- Applying CHEST THRUSTS. An alternate technique to the abdominal thrust is the chest thrust. This technique is useful when the casualty has an abdominal wound, when the casualty is pregnant, or when the casualty is so large that you cannot wrap your arms around the abdomen. TO apply chest thrusts with casualty sitting or standing:
  - o Stand behind the casualty and wrap your arms around his chest with your arms under his armpits.
  - o Make a fist with one hand and place the thumb side of the fist in the middle of the breastbone (take care to avoid the tip of the breastbone and the margins of the ribs).
  - o Grasp the fist with the other hand and exert thrusts (Figure 2-21).
  - o Each thrust should be delivered slowly, distinctly, and with the intent of relieving the obstruction.
  - o Perform chest thrusts until the obstruction is expelled or the casualty becomes unconscious.

    o   If the casualty becomes unconscious, call for help as you proceed with steps to open the airway and perform rescue breathing. (See task 081-831-1042, Perform Mouth-to-Mouth Resuscitation.)

2-14. Open an Obstructed Airway—Casualty Lying or Unconscious (081-831-1042)
The following procedures are used to expel an airway obstruction in a casualty who is lying down, who becomes unconscious, or is found unconscious (the cause unknown):

- If a casualty who is choking becomes unconscious, call for help, open the airway, perform a finger sweep, and attempt rescue breathing (paragraphs 2-2 through 2-4). If you still cannot administer rescue breathing due to an airway blockage, then remove the airway obstruction using the procedures in steps a through e below.
- If a casualty is unconscious when you find him (the cause unknown), assess or evaluate the situation, call for help, position the casualty on his back, open the airway, establish breathlessness, and attempt to perform rescue breathing (paragraphs 2-2 through 2-8).

Figure 2-21. Profile view of chest thrust

Figure 2-22: Abdominal thrust on unconscious casualty

   a. Open the airway and attempt rescue breathing. (See task 081-831-1042, Perform Mouth-to-Mouth Resuscitation.)

   b. If still unable to ventilate the casualty, perform 6 to 10manual (abdominal or chest) thrusts. (Note that the abdominal thrusts are used when casualty does not have abdominal wounds; is not pregnant or extremely overweight.) To perform the abdominal thrusts:

      (1) Kneel astride the casualty's thighs (Figure 2-22).

      (2) Place the heel of one hand against the casualty's abdomen (in the midline slightly above the navel but well below the tip of the breastbone). Place your other hand on top of the first one. Point your fingers toward the casualty's head.

      (3) Press into the casualty's abdomen with a quick, forward and upward thrust. You can use your body weight to perform the maneuver. Deliver each thrust slowly and distinctly.

      (4) Repeat the sequence of abdominal thrusts, finger sweep, and rescue breathing (attempt to ventilate) as long as necessary to remove the object from the obstructed airway. See paragraph d below.

      (5) If the casualty's chest rises, proceed to feeling for pulse.

   c. Apply chest thrusts. (Note that the chest thrust technique is an alternate method that is used when the casualty has an abdominal wound, when the casualty is so large that you cannot wrap your arms around the abdomen, or when the casualty is pregnant.) To perform the chest thrusts:

(1) Place the unconscious casualty on his back, face up, and open his mouth. Kneel close to the side of the casualty's body.

- Locate the lower edge of the casualty's ribs with your fingers. Run the fingers up along the rib cage to the notch (Figure2-23A).
- Place the middle finger on the notch and the index finger next to the middle finger on the lower edge of the breastbone. Place the heel of the other hand on the lower half of the breastbone next to the two fingers (Figure 2-23B).
- Remove the fingers from the notch and place that hand on top of the positioned hand on the breastbone, extending or interlocking the fingers (Figure 2-23C).
- Straighten and lock your elbows with your shoulders directly above your hands without bending the elbows, rocking, or allowing the shoulders to sag. Apply

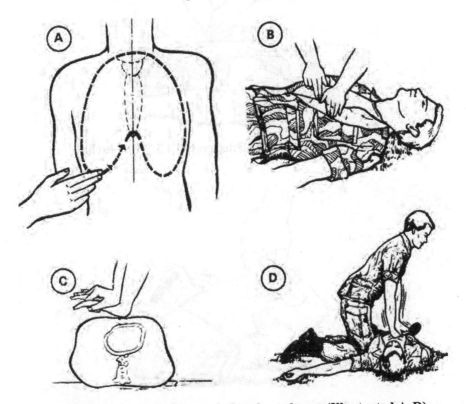

Figure 2-23: Hand placement for chest thrust (Illustrated A-D)

enough pressure to depress the breastbone 1 1/2 to 2 inches, then release the pressure completely (Figure 2-23D). Do this 6 to 10 times. Each thrust should be delivered slowly and distinctly. See Figure 2-24 for another view of the breastbone being depressed.

(2) Repeat the sequence of chest thrust, finger sweep, and rescue breathing as long as necessary to clear the object from the obstructed airway. See paragraph d below.

(3) If the casualty's chest rises, proceed to feeling for his pulse.

d. Finger Sweep. If you still cannot administer rescue breathing due to an airway obstruction, then remove the airway obstruction using the procedures in steps (1) and (2) below.

(1) Place the casualty on his back, face up, turn the unconscious casualty as a unit, and call out for help.

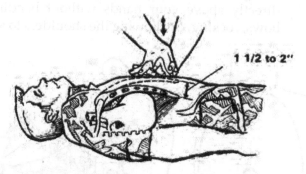

1 1/2 to 2"

Figure 2-24: Breastbone depressed 1 1/2 to 2 inches

Figure 2-25: Opening casualty's mouth (tongue-jaw lift)

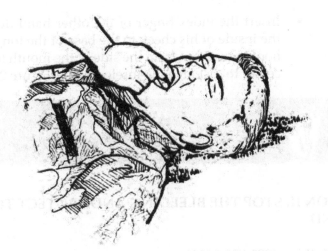

Figure 2-26: Opening casualty's mouth (crossed-finger method)

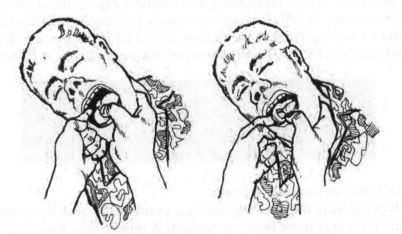

Figure 2-27: Using finger to dislodge foreign body

(2) Perform finger sweep, keep casualty face up, use tongue-jaw lift to open mouth.
- Open the casualty's mouth by grasping both his tongue and lower jaw between your thumb and fingers and lifting (tongue-jaw lift) (Figure 2-25). If you are unable to open his mouth, cross your fingers and thumb (crossed-finger method) and push his teeth apart (Figure 2-26) by pressing your thumb against his upper teeth and pressing your finger against his lower teeth.

- Insert the index finger of the other hand down along the inside of his cheek to the base of the tongue. Use a hooking motion from the side of the mouth toward the center to dislodge the foreign body (Figure 2-27).

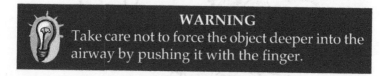

WARNING
Take care not to force the object deeper into the airway by pushing it with the finger.

SECTION II. STOP THE BLEEDING AND PROTECT THE WOUND

2-15. Clothing (081-831-1016)

In evaluating the casualty for location, type, and size of the wound or injury, cut or tear his clothing and carefully expose the entire area of the wound. This procedure is necessary to avoid further contamination, Clothing stuck to the wound should be left in place to avoid further injury. DO NOT touch the wound; keep it as clean as possible.

WARNING (081-831-1016)
DO NOT REMOVE protective clothing in a chemical environment. Apply dressings over the protective clothing.

2-16. Entrance and Exit Wounds

Before applying the dressing, carefully examine the casualty to determine if there is more than one wound. A missile may have entered atone point and exited at another point. The EXIT wound is usually LARGER than the entrance wound.

WARNING
Casualty should be continually monitored for development of conditions which may require the performance of necessary basic lifesaving measures, such as clearing the airway and mouth-to-mouth resuscitation. All open (or penetrating) wounds should be checked for a point of entry and exit and treated accordingly.

> **WARNING**
> If the missile lodges in the body (fails to exit), DO NOT attempt to remove it or probe the wound. Apply a dressing. If there is an object extending from (impaled in) the wound, DO NOT remove the object. Apply a dressing around the object and use additional improvised bulky materials dressings (use the cleanest material available) to build up the area around the object. Apply a supporting bandage over the bulky materials to hold them in place.

2-17. Field Dressing (081-831-1016)

a. Use the casualty's field dressing; remove it from the wrapper and grasp the tails of the dressing with both hands (Figure 2-28).

> **WARNING**
> DO NOT touch the white (sterile) side of the dressing, and DO NOT allow the white (sterile) side of the dressing to come in contact with any surface other than the wound.

b. Hold the dressing directly over the wound with the white side Down. Pull the dressing open (Figure 2-29) and place it directly over the wound (Figure 2-30).

c. Hold the dressing in place with one hand. Use the other hand to wrap one of the tails around the injured part, covering about

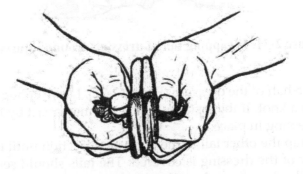

Figure 2-28: Grasping tails of dressing with both hands

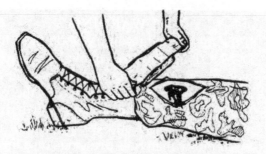

Figure 2-29: Pulling dressing open

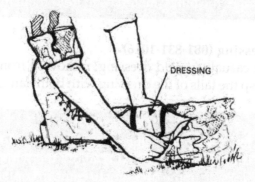

Figure 2-30: Placing dressing directly on wound

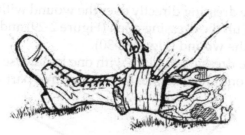

Figure 2-31: Wrapping tail of dressing around injured part

one-half of the dressing (Figure 2-31). Leave enough of the tail for a knot. If the casualty is able, he may assist by holding the dressing in place.

d. Wrap the other tail in the opposite direction until the remainder of the dressing is covered. The tails should seal the sides of the dressing to keep foreign material from getting under it.

e. Tie the tails into a non-slip knot over the outer edge of the dressing (Figure 2-32). DO NOT TIE THE KNOT OVER THE WOUND. In order to allow blood to flow to the rest of an injured limb, tie the dressing firmly enough to prevent it from slipping but without causing a tourniquet-like effect; that is, the skin beyond the injury becomes cool, blue, or numb.

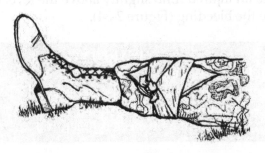

Figure 2-32: Tails tied into nonslip knot

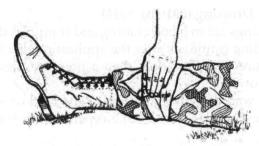

Figure 2-33: Direct manual pressure applied

Figure 2-34: Injured limb elevated

2-18. Manual Pressure (081-831-1016)

a. If bleeding continues after applying the sterile field dressing, direct manual pressure may be used to help control bleeding. Apply such pressure by placing a hand on the dressing and exerting firm pressure for5 to 10 minutes (Figure 2-33). The casualty may be asked to do this himself if he is conscious and can follow instructions.

b. Elevate an injured limb slightly above the level of the heart to reduce the bleeding (Figure 2-34).

> **WARNING**
>
> DO NOT elevate a suspected fractured limb unless it has been properly splinted. (To splint a fracture before elevating, see task081-831-1034, Splint a Suspected Fracture.)

c. If the bleeding stops, check and treat for shock. If the bleeding continues, apply a pressure dressing.

2-19. Pressure Dressing (081-831-1016)

Pressure dressings aid in blood clotting and compress the open blood vessel. If bleeding continues after the application of a field dressing, manual pressure, and elevation, then a pressure dressing must be applied as follows:

a. Place a wad of padding on top of the field dressing, directly over the wound (Figure 2-35). Keep injured extremity elevated.

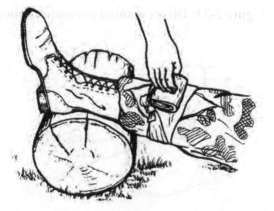

Figure 2-35: Wad of padding on top of field dressing

✍ **NOTE**

Improvised bandages may be made from strips of cloth. These strips may be made from T-shirts, socks, or other garments.

---

b. Place an improvised dressing (or cravat, if available) over the wad of padding (Figure 2-36). Wrap the ends tightly around the injured limb, covering the previously placed field dressing (Figure 2-37).

c. Tie the ends together in a non-slip knot, directly over the wound site (Figure 2-38). DO NOT tie so tightly that it has a tourniquet-like effect. If bleeding continues and all other measures have failed, or if the limb is severed, then apply a

Figure 2-36: Improvised dressing over wad of padding

Figure 2-37: Ends of improvised dressing wrapped tightly around limb

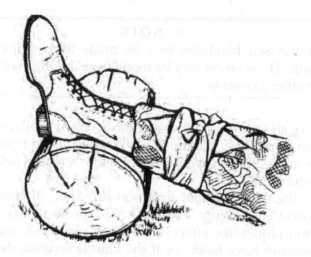

Figure 2-38: Ends of improvised dressing tied together in nonslip knot

tourniquet. Use the tourniquet as a LAST RESORT. When the bleeding stops, check and treat for shock.

✍ **NOTE**
Wounded extremities should be checked periodically for adequate circulation. The dressing must be loosened if the extremity becomes cool, blue or gray, or numb.

✍ * **NOTE**
If bleeding continues and all other measures have failed (dressing and covering wound, applying direct manual pressure, elevating limb above heart level, and applying pressure dressing maintaining limb elevation), then apply digital pressure. See Appendix E for appropriate pressure points.

2-20. Tourniquet (081-831-1017)

A tourniquet is a constricting band placed around an arm or leg to control bleeding. A soldier whose arm or leg has been completely amputated may not be bleeding when first discovered, but a tourniquet should be applied anyway. This absence of bleeding is due to the body's normal defenses (contraction of blood vessels) as a result of the amputation, but after a period of time bleeding will start as the blood vessels relax. Bleeding from a major artery of the thigh, lower leg, or arm and bleeding from multiple arteries (which occurs in a traumatic amputation) may prove to be beyond control by manual pressure. If the pressure dressing under firm hand pressure becomes soaked with blood and the wound continues to bleed, apply a tourniquet.

> **WARNING**
> Casualty should be continually monitored for development of conditions which may require the performance of necessary basic life-saving measures, such as: clearing the airway, performing mouth-to-mouth resuscitation, preventing shock, and/or bleeding control. All open (or penetrating) wounds should be checked for a point of entry or exit and treated accordingly.

* The tourniquet should not be used unless a pressure dressing has failed to stop the bleeding or an arm or leg has been cut off. On occasion, tourniquets have injured blood vessels and nerves. If left in place too long, a tourniquet can cause loss of an arm or leg. Once applied, it must stay in place, and the casualty must be taken to the nearest medical treatment facility as soon as possible. DO NOT loosen or release a tourniquet after it has been applied and the bleeding has stopped.

    a. Improvising a Tourniquet (081-831-1017). In the absence of a specially designed tourniquet, a tourniquet may be made from a strong, pliable material, such as gauze or muslin bandages, clothing, or kerchiefs. An improvised tourniquet is used with a rigid stick-like object. To minimize skin damage, ensure that the improvised tourniquet is at least 2 inches wide.

> **WARNING**
> The tourniquet must be easily identified or easily seen.

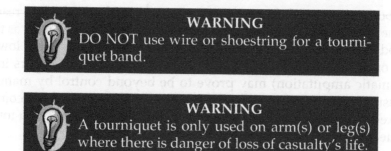

WARNING

DO NOT use wire or shoestring for a tourniquet band.

WARNING

A tourniquet is only used on arm(s) or leg(s) where there is danger of loss of casualty's life.

b. Placing the Improvised Tourniquet (081-831-1017).

(1) Place the tourniquet around the limb, between the wound and the body trunk (or between the wound and the heart). Place the tourniquet 2 to 4 inches from the edge of the wound site (Figure 2-39). Never place it directly over a wound or fracture or directly on a joint (wrist, elbow, or knee). For wounds just below a joint, place the tourniquet just above and as close to the joint as possible.

(2) The tourniquet should have padding underneath. If possible, place the tourniquet over the smoothed sleeve or trouser leg to prevent the skin from being pinched or twisted. If the tourniquet is long enough, wrap it around the limb several times, keeping the material as flat as possible. Damaging the skin may deprive the surgeon of skin required to cover an amputation. Protection of the skin also reduces pain.

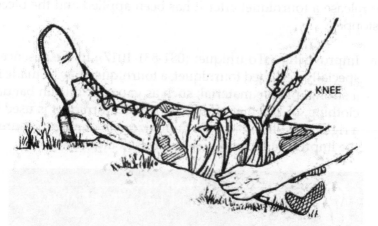

Figure 2-39: Tourniquet 2 to 4 inches above wound

c. Applying the Tourniquet (081-831-1017).
   (1) Tie a half-knot. (A half-knot is the same as the first part of tying a shoe lace.)
   (2) Place a stick (or similar rigid object) on top of the half-knot (Figure 2-40).
   (3) Tie a full knot over the stick (Figure 2-41).
   (4) Twist the stick (Figure 2-42) until the tourniquet is tight around the limb and/or the bright red bleeding has stopped. In the case of amputation, dark oozing blood may continue for a short time. This is the blood trapped in the area between the wound and tourniquet.
   (5) Fasten the tourniquet to the limb by looping the free ends of the tourniquet over the ends of the stick. Then bring the ends around the limb to prevent the stick from loosening. Tie them together under the limb (Figure 2-43A and B).

Figure 2-40: Rigid object on top of half-knot

Figure 2-41: Full knot over rigid object

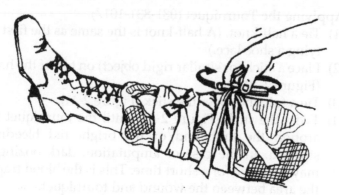

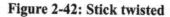

Figure 2-42: Stick twisted

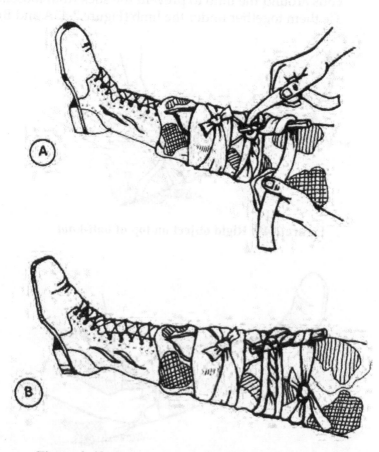

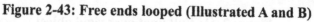

Figure 2-43: Free ends looped (Illustrated A and B)

> ✍ NOTE (081-831-1017)
> Other methods of securing the stick may be used as long as the stick does not unwind and no further injury results.

> ✍ NOTE
> If possible, save and transport any severed (amputated) limbs or body parts with (but out of sight of) the casualty.

(6) DO NOT cover the tourniquet–you should leave it in full view. If the limb is missing (total amputation), apply a dressing to the stump.

(7) Mark the casualty's forehead, if possible, with a "T" to indicate a tourniquet has been applied. If necessary, use the casualty's blood to make this mark.

(8) Check and treat for shock.

(9) Seek medical aid.

> ⚠ CAUTION (081-831-1017)
> DO NOT LOOSEN OR RELEASE THE TOURNIQUET ONCE IT HAS BEEN APPLIED BECAUSE IT COULD ENHANCE THE PROBABILITY OF SHOCK.

SECTION III. CHECK AND TREAT FOR SHOCK

2-21. Causes and Effects

a. Shock may be caused by severe or minor trauma to the body. It usually is the result of—
- Significant loss of blood.
- Heart failure.
- Dehydration.
- Severe and painful blows to the body.
- Severe burns of the body.
- Severe wound infections.
- Severe allergic reactions to drugs, foods, insect stings, and snakebites.

b. Shock stuns and weakens the body. When the normal blood flow in the body is upset, death can result. Early identification and proper treatment may save the casualty's life.

c. See FM 8-230 for further information and details on specific types of shock and treatment.

2-22. Signs/Symptoms (081-831-1000)

Examine the casualty to see if he has any of the following signs/ symptoms:

- Sweaty but cool skin (clammy skin).
- Paleness of skin.
- Restlessness, nervousness.
- Thirst.
- Loss of blood (bleeding).
- Confusion (or loss of awareness).
- Faster-than-normal breathing rate.
- Blotchy or bluish skin (especially around the mouth and lips).
- Nausea and/or vomiting.

2-23. Treatment/Prevention (081-831-1005)

In the field, the procedures to treat shock are identical to procedures that would be performed to prevent shock. When treating a casualty, assume that shock is present or will occur shortly. By waiting until actual signs/symptoms of shock are noticeable, the rescuer may jeopardize the casualty's life.

a. Position the Casualty. (DO NOT move the casualty or his limbs if suspected fractures have not been splinted. See Chapter 4 for details.)

(1) Move the casualty to cover, if cover is available and the situation permits.

(2) Lay the casualty on his back.

✍ NOTE

A casualty in shock after suffering a heart attack, chest wound, or breathing difficulty, may breathe easier in a sitting position. If this is the case, allow him to sit upright, but monitor carefully in case his condition worsens.

---

(3) Elevate the casualty's feet higher than the level of his heart. Use a stable object (a box, field pack, or rolled up clothing) so that his feet will not slip off (Figure 2-44).

Figure 2-44: Clothing loosened and feet elevated

WARNING
DO NOT elevate legs if the casualty has an unsplinted broken leg, head injury, or abdominal injury. (See task 081-831-1034, Splint a Suspected Fracture, and task 081-831-1025, Apply a Dressing to an Open Abdominal Wound.)

WARNING (081-831-1005)
Check casualty for leg fracture(s) and splint, if necessary, before elevating his feet. For a casualty with an abdominal wound, place knees in an upright (flexed) position.

(4) Loosen clothing at the neck, waist, or wherever it may be binding.

⚠ **CAUTION (081-831-1005)**
DO NOT LOOSEN OR REMOVE protective clothing in a chemical environment.

(5) Prevent chilling or overheating. The key is to maintain body temperature. In cold weather, place a blanket or other like item over him to keep him warm and under him to prevent chilling (Figure 2-45). However, if a tourniquet has been applied, leave it exposed (if possible). In hot weather, place the casualty in the shade and avoid excessive covering.

Figure 2-45: Body temperature maintained

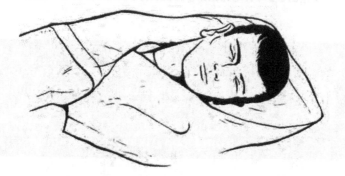

Figure 2-46: Casualty's head turned to side

(6) Calm the casualty. Throughout the entire procedure of treating and caring for a casualty, the rescuer should reassure the casualty and keep him calm. This can be done by being authoritative (taking charge) and by showing self-confidence. Assure the casualty that you are there to help him.

(7) Seek medical aid.

b. Food and/or Drink. During the treatment/prevention of shock, DO NOT give the casualty any food or drink. If you must leave the casualty or if he is unconscious, turn his head to the side to prevent him from choking should he vomit (Figure 2-46).

c. Evaluate Casualty. If necessary, continue with the casualty's evaluation.

CHAPTER 3

FIRST AID FOR SPECIAL WOUNDS

* Basic lifesaving steps are discussed in Chapters 1 and 2: clear the airway/restore breathing, stop the bleeding, protect the wound, and treat/prevent shock. They apply to first aid measures for all injuries. Certain types of wounds and burns will require special precautions and procedures when applying these measures. This chapter discusses first aid procedures for special wounds of the head, face, and neck; chest and stomach wounds; and burns. It also discusses the techniques for applying dressings and bandages to specific parts of the body.

SECTION I. GIVE PROPER FIRST AID FOR HEAD INJURIES

3-1. Head Injuries
A head injury may consist of one or a combination of the following conditions: a concussion, a cut or bruise of the scalp, or a fracture of the skull with injury to the brain and the blood vessels of the scalp. The damage can range from a minor cut on the scalp to a severe brain injury which rapidly causes death. Most head injuries lie somewhere between the two extremes. Usually, serious skull fractures and brain injuries occur together; however, it is possible to receive a serious brain injury without a skull fracture. The brain is a very delicate organ; when it is injured, the casualty may vomit, become sleepy, suffer paralysis, or lose consciousness and slip into a coma. All severe head injuries are potentially life-threatening. For recovery and return to normal function, casualties require proper first aid as a vital first step.

3-2. Signs/Symptoms (081-831-1000)
A head injury may be open or closed. In open injuries, there is a visible wound and, at times, the brain may actually be seen. In closed injuries, no visible injury is seen, but the casualty may experience the

same signs and symptoms. Either closed or open head injuries can be life-threatening if the injury has been severe enough; thus, if you suspect a head injury, evaluate the casualty for the following:

- Current or recent unconsciousness (loss of consciousness).
- Nausea or vomiting.
- Convulsions or twitches (involuntary jerking and shaking).
- Slurred speech.
- Confusion.
- Sleepiness (drowsiness).
- Loss of memory (does casualty know his own name, where he is, and so forth).
- Clear or bloody fluid leaking from nose or ears.
- Staggering in walking.
- Dizziness.
- A change in pulse rate.
- Breathing problems.
- Eye (vision) problems, such as unequal pupils.
- Paralysis.
- Headache.
- Black eyes.
- Bleeding from scalp/head area.
- Deformity of the head.

3-3. General First Aid Measures (081-831-1000)

a. General Considerations. The casualty with a head injury (or suspected head injury) should be continually monitored for the development of conditions which may require the performance of the necessary basic lifesaving measures, therefore be prepared to—
- Clear the airway (and be prepared to perform the basic lifesaving measures). Treat as a suspected neck/spinal injury until proven otherwise. (See Chapter 4 for more information.)
- Place a dressing over the wounded area. DO NOT attempt to clean the wound.
- Seek medical aid.
- Keep the casualty warm.
- DO NOT attempt to remove a protruding object from the head.
- DO NOT give the casualty anything to eat or drink.

b. Care of the Unconscious Casualty. If a casualty is unconscious as the result of a head injury, he is not able to defend himself. He may lose his sensitivity to pain or ability to cough up blood or mucus that may be plugging his airway. An unconscious casualty must be evaluated for breathing difficulties, uncontrollable bleeding, and spinal injury.

(1) Breathing. The brain requires a constant supply of oxygen. A bluish (or in an individual with dark skin—grayish) color of skin around the lips and nail beds indicates that the casualty is not receiving enough air (oxygen). Immediate action must be taken to clear the airway, to position the casualty on his side, or to give artificial respiration. Be prepared to give artificial respiration if breathing should stop.

(2) Bleeding. Bleeding from a head injury usually comes from blood vessels within the scalp. Bleeding can also develop inside the skull or within the brain. In most instances bleeding from the head can be controlled by proper application of the field first aid dressing.

⚠ CAUTION (081-831-1033)

DO NOT attempt to put unnecessary pressure on the wound or attempt to push any/brain matter back into the head (skull). DO NOT apply a pressure dressing.

(3) Spinal injury. A person that has an injury above the collar bone or a head injury resulting in an unconscious state should be suspected of having a neck or head injury with spinal cord damage. Spinal cord injury may be indicated by a lack of responses to stimuli, stomach distention (enlargement), or penile erection.

(a) Lack of responses to stimuli. Starting with the feet, use a sharp pointed object–a sharp stick or something similar, and prick the casualty lightly while observing his face. If the casualty blinks or frowns, this indicates that he has feeling and may not have an injury to the spinal cord. If you observe no response in the casualty's reflexes after pricking upwards toward the chest region, you must use extreme caution and treat the casualty for an injured spinal cord.

(b) Stomach distention (enlargement). Observe the casualty's chest and stomach. If the stomach is distended (enlarged) when the casualty takes a breath and the chest moves slightly, the casualty may have a spinal injury and must be treated accordingly.

(c) Penile erection. A male casualty may have a penile erection, an indication of a spinal injury.

⚠ CAUTION

Remember to suspect any casualty who has a severe head injury or who is/unconscious as possibly having a broken neck or a spinal cord injury! It is better to treat conservatively and assume that the neck/spinal cord is injured rather than to chance further injuring the casualty. Consider this when you position the casualty. See Chapter 4, paragraph 4-9 for treatment procedures of spinal column injuries.

---

c. Concussion. If an individual receives a heavy blow to the head or face, he may suffer a brain concussion, which is an injury to the brain that involves a temporary loss of some or all of the brain's ability to function. For example, the casualty may not breathe properly for a short period of time, or he may become confused and stagger when he attempts to walk. A concussion may only last for a short period of time. However, if a casualty is suspected of having suffered a concussion, he must be seen by a physician as soon as conditions permit.

d. Convulsions. Convulsions (seizures/involuntary jerking) may occur after a mild head injury. When a casualty is convulsing, protect him from hurting himself. Take the following measures:

(1) Ease him to the ground.

(2) Support his head and neck.

(3) Maintain his airway.

(4) Call for assistance.

(5) Treat the casualty's wounds and evacuate him immediately.

e. Brain Damage. In severe head injuries where brain tissue is protruding, leave the wound alone; carefully place a first aid dressing over the tissue. DO NOT remove or disturb any foreign matter that maybe in the wound. Position the casualty so that his head is higher than his body. Keep him warm and seek medical aid immediately.

✍ **NOTE**

- DO NOT forcefully hold the arms and legs if they are jerking because this can lead to broken bones.
- DO NOT force anything between the casualty's teeth-especially if they are tightly clenched because this may obstruct the casualty's airway.
- Maintain the casualty's airway if necessary.

3-4. Dressings and Bandages (081-831-1000 and 081-831-1033)

* a. Evaluate the Casualty (081-831-1000). Be prepared to perform lifesaving measures. The basic lifesaving measures may include clearing the airway, rescue breathing, treatment for shock, and/or bleeding control.

b. Check Level of Consciousness/Responsiveness (081-831-1033). With a head injury, an important area to evaluate is the casualty's level of consciousness and responsiveness. Ask the casualty questions such as—

- "What is your name?" (Person)
- "Where are you?" (Place)
- "What day/month/year is it?" (Time)

Any incorrect responses, inability to answer, or changes in responses should be reported to medical personnel. Check the casualty's level of consciousness every 15 minutes and note any changes from earlier observations.

c. Position the Casualty (081-831-1033).

WARNING (081-831-1033)

DO NOT move the casualty if you suspect he has sustained a neck, spine, or severe, head injury (which produces any signs or symptoms other than minor bleeding). See task 081-831-1000, Evaluate the Casualty.

- If the casualty is conscious or has a minor (superficial) scalp wound:
  - ○ Have the casualty sit up (unless other injuries prohibit or he is unable); OR
  - ○ If the casualty is lying down and is not accumulating fluids or drainage in his throat, elevate his head slightly; OR
  - ○ If the casualty is bleeding from or into his mouth or throat, turn his head to the side or position him on his side so that the airway will be clear. Avoid pressure on the wound or place him on his side—opposite the site of the injury (Figure 3-1).
- If the casualty is unconscious or has a severe head injury, then suspect and treat him as having a potential neck or spinal injury, immobilize and DO NOT move the casualty.

✍ NOTE (081-831-1033)

If the casualty is choking and/or vomiting or is bleeding from or into his mouth (thus compromising his airway), position him on his side so that his airway will be clear. Avoid pressure on the wound; place him on his side opposite the side of the injury.

WARNING (081-831-1033)

If it is necessary to turn a casualty with a suspected neck/spine injury; roll the casualty gently onto his side, keeping the head, neck, and body aligned while providing support for the head and neck. DO NOT roll the casualty by yourself but seek assistance. Move him only if absolutely necessary, otherwise keep the casualty immobilized to prevent further damage to the neck/spine.

---

d. Expose the Wound (081-831-1033).
- Remove the casualty's helmet (if necessary).
- In a chemical environment:
  - ○ If mask and/or hood is not breached, apply no dressing to the head wound casualty. If the "all clear" has

not been given, DO NOT remove the casualty's mask to attend the head wound: OR
- o If mask and/or hood have been breached and the "all clear" has not been given, try to repair the breach with tape and apply no dressing; OR
- o If mask and/or hood have been breached and the "all clear" has been given the mask can be removed and a dressing applied.

WARNING

DO NOT attempt to clean the wound, or remove a protruding object.

✍ **NOTE**

If there is an object extending from the wound, DO NOT remove the object. Improvise/bulky dressings from the cleanest material available and place these dressings around the protruding object for support after applying the field dressing.

✍ **NOTE**

Always use the casualty's field dressing, not your own!

---

e. Apply a Dressing to a Wound of the Forehead/Back of Head (081-831-1033). To apply a dressing to a wound of the forehead or back of the head—
(1) Remove the dressing from the wrapper.
(2) Grasp the tails of the dressing in both hands.

Figure 3-1: Casualty lying on side opposite injury

(3) Hold the dressing (white side down) directly over the wound. DO NOT touch the white (sterile) side of the dressing or allow anything except the wound to come in contact with the white side.

(4) Place it directly over the wound.

(5) Hold it in place with one hand. If the casualty is able, he may assist.

(6) Wrap the first tail horizontally around the head; ensure the tail covers the dressing (Figure 3-2).

(7) Hold the first tail in place and wrap the second tail in the opposite direction, covering the dressing (Figure 3-3).

(8) Tie a nonslip knot and secure the tails at the side of the head, making sure they DO NOT cover the eyes or ears (Figure 3-4).

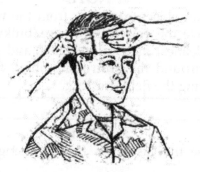

Figure 3-2: First tail of dressing wrapped horizontally around head

Figure 3-3: Second tail wrapped in opposite direction

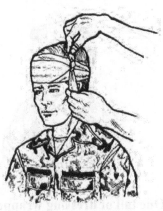

Figure 3-4: Tails tied in nonslip knot at side of head

f. Apply a Dressing to a Wound on Top of the Head (081-831-1033). To apply a dressing to a wound on top of the head–
    (1) Remove the dressing from the wrapper.
    (2) Grasp the tails of the dressing in both hands.
    (3) Hold it (white side down) directly over the wound.
    (4) Place it over the wound (Figure 3-5).
    (5) Hold it in place with one hand. If the casualty is able, he may assist.
    (6) Wrap one tail down under the chin (Figure 3-6), up in front of the ear, over the dressing, and in front of the other ear.

Figure 3-5: Dressing placed over wound

Figure 3-6: One tail of dressing wrapped under chin

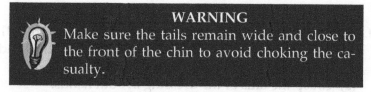

WARNING

Make sure the tails remain wide and close to the front of the chin to avoid choking the casualty.

(7) Wrap the remaining tail under the chin in the opposite direction and up the side of the face to meet the first tail (Figure 3-7).

(8) Cross the tails (Figure 3-8), bringing one around the forehead (above the eyebrows) and the other around the back of the head(at the base of the skull) to a point just above and in front of the opposite ear, and tie them using a non slip knot (Figure 3-9).

Figure 3-7: Remaining tail wrapped under chin in opposite direction

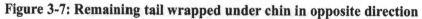

Figure 3-8: Tails of dressing crossed with one around forehead

Figure 3-9: Tails tied in nonslip knot (in front of and above ear)

g. Apply a Triangular Bandage to the Head. To apply a triangular bandage to the head–
   (1) Turn the base (longest side) of the bandage up and center its base on center of the forehead, letting the point (apex) fall on the back of the neck (Figure 3-10 A).
   (2) Take the ends behind the head and cross the ends over the apex.
   (3) Take them over the forehead and tie them (Figure 3-10 B).
   (4) Tuck the apex behind the crossed part of the bandage and/or secure it with a safety pin, if available (Figure 3-10 C).

Figure 3-10: Triangular bandage applied to head (Illustrated A thru C)

Figure 3-11: Cravat bandage applied to head (Illustrated A thru C)

h. Apply a Cravat Bandage to the Head. To apply a cravat bandage to the head–
(1) Place the middle of the bandage over the dressing (Figure
3-11 A).
(2) Cross the two ends of the bandage in opposite directions
completely around the head (Figure 3-11 B).
(3) Tie the ends over the dressing (Figure 3-11 C).

SECTION II. GIVE PROPER FIRST AID FOR FACE AND NECK INJURIES

3-5. Face Injuries
Soft tissue injuries of the face and scalp are common. Abrasions
(scrapes) of the skin cause no serious problems. Contusions (injury
without a break in the skin) usually cause swelling. A contusion of

the scalp looks and feels like a lump. Laceration (cut) and avulsion (torn away tissue) injuries are also common. Avulsions are frequently caused when a sharp blow separates the scalp from the skull beneath it. Because the face and scalp are richly supplied with blood vessels (arteries and veins), wounds of these areas usually bleed heavily.

3-6. Neck Injuries

Neck injuries may result in heavy bleeding. Apply manual pressure above and below the injury and attempt to control the bleeding. Apply a dressing. Always evaluate the casualty for a possible neck fracture/ spinal cord injury; if suspected, seek medical treatment immediately.

✍ *NOTE

Establish and maintain the airway in cases of facial or neck injuries. If a neck fracture or spinal cord injury is suspected, immobilize or stabilize casualty. See Chapter 4 for further information on treatment of spinal injuries.

3-7. Procedure

When a casualty has a face or neck injury, perform the measures below.

    a. Step ONE. Clear the airway. Be prepared to perform any of the basic lifesaving steps. Clear the casualty's airway (mouth) with your fingers, remove any blood, mucus, pieces of broken teeth or bone, or bits of flesh, as well as any dentures.

    b. Step TWO. Control any bleeding, especially bleeding that obstructs the airway. Do this by applying direct pressure over a first aid dressing or by applying pressure at specific pressure points on the face, scalp, or temple. (See Appendix E for further information on pressure points.) If the casualty is bleeding from the mouth, position him as indicated (c below) and apply manual pressure.

⚠ CAUTION

Take care not to apply too much pressure to the scalp if a skull fracture is suspected.

---

    c. Step THREE. Position the casualty. If the casualty is bleeding from the mouth (or has other drainage, such as mucus,

Figure 3-12: Casualty leaning forward to permit drainage

vomit, or so forth) and is conscious, place him in a comfortable sitting position and have him lean forward with his head tilted slightly down to permit free drainage (Figure 3-12). DO NOT use the sitting position if–

- It would be harmful to the casualty because of other injuries.
- The casualty is unconscious, in which case, place him on his side (Figure 3-13). If there is a suspected injury to the neck or spine, immobilize the head before turning the casualty on his side.

△ CAUTION

If you suspect the casualty has a neck/spinal injury, then immobilize his head/neck and treat him as outlined in Chapter 4.

---

d. Step FOUR. Perform other measures.
   (1) Apply dressings/bandages to specific areas of the face.
   (2) Check for missing teeth and pieces of tissue. Check for detached teeth in the airway. Place detached teeth, pieces of ear or nose on a field dressing and send them along with the casualty to the medical facility. Detached teeth should be kept damp.
   (3) Treat for shock and seek medical treatment IMMEDIATELY.

WOUND

Figure 3-13: Casualty lying on side

3-8. Dressings and Bandages (081-831-1033)

a. Eye Injuries. The eye is a vital sensory organ, and blindness is a severe physical handicap. Timely first aid of the eye not only relieves pain but also helps prevent shock, permanent eye injury, and possible loss of vision. Because the eye is very sensitive, any injury can be easily aggravated if it is improperly handled. Injuries of the eye may be quite severe. Cuts of the eyelids can appear to be very serious, but if the eyeball is not involved, a person's vision usually will not be damaged. However, lacerations (cuts) of the eyeball can cause permanent damage or loss of sight.

(1) Lacerated/torn eyelids. Lacerated eyelids may bleed heavily, but bleeding usually stops quickly. Cover the injured eye with a sterile dressing. DO NOT put pressure on the wound because you may injure the eyeball. Handle torn eyelids very carefully to prevent further injury. Place any detached pieces of the eyelid on a clean bandage or dressing and immediately send them with the casualty to the medical facility.

(2) Lacerated eyeball (injury to the globe). Lacerations or cuts to the eyeball may cause serious and permanent eye damage. Cover the injury with a loose sterile dressing. DO NOT put pressure on the eyeball because additional damage may occur. An important point to remember is that when one eyeball is injured, you should immobilize both eyes. This is done by applying a bandage to both eyes. Because the eyes move together, covering both will lessen the chances of further damage to the injured eye.

> **⚠ CAUTION**
> DO NOT apply pressure when there is a possible laceration of the eyeball. The eyeball contains fluid. Pressure applied over the eye will force the fluid out, resulting in/permanent injury. APPLY PROTECTIVE DRESSING WITHOUT ADDED PRESSURE.

(3) Extruded eyeballs. Soldiers may encounter casualties with severe eye injuries that include an extruded eyeball (eyeball out-of-socket). In such instances you should gently cover the extruded eye with a loose moistened dressing and also cover the unaffected eye. DO NOT bind or exert pressure on the injured eye while applying a loose dressing. Keep the casualty quiet, place him on his back, treat for shock (make warm and comfortable), and evacuate him immediately.

(4) Burns of the eyes. Chemical burns, thermal (heat) burns, and light burns can affect the eyes.

　(a) Chemical burns. Injuries from chemical burns require immediate first aid. Chemical burns are caused mainly by acids or alkalies. The first aid is to flush the eye(s) immediately with large amounts of water for at least 5 to 20 minutes, or as long as necessary to flush out the chemical. If the burn is an acid burn, you should flush the eye for at least 5 to 10 minutes. If the burn is an alkali burn, you should flush the eye for at least 20 minutes. After the eye has been flushed, apply a bandage over the eyes and evacuate the casualty immediately.

　(b) Thermal burns. When an individual suffers burns of the face from a fire, the eyes will close quickly due to extreme heat. This reaction is a natural reflex to protect the eyeballs; however, the eyelids remain exposed and are frequently burned. If a casualty receives burns of the eyelids/face, DO NOT apply a dressing; DO NOT TOUCH; seek medical treatment immediately.

　(c) Light burns. Exposure to intense light can burn an individual. Infrared rays, eclipse light (if the casualty has looked directly at the sun), or laser burns cause injuries of the exposed eyeball. Ultraviolet rays from arc welding can cause a superficial burn to the surface of the eye. These injuries are generally not painful but may cause permanent damage to the eyes. Immediate first aid is usually not required. Loosely bandaging the eyes may make the casualty more comfortable and protect his eyes from further injury caused by exposure to other bright lights or sunlight.

⚠ CAUTION

In certain instances both eyes are usually bandaged; but, in hazardous surroundings leave the uninjured eye uncovered so that the casualty may be able to see.

---

b. Side-of-Head or Cheek Wound (081-831-1033).
Facial injuries to the side of the head or the cheek may bleed profusely (Figure 3-14). Prompt action is necessary to ensure that the airway remains open and also to control the bleeding. It may be necessary to apply a dressing. To apply a dressing—
(1) Remove the dressing from its wrapper.
(2) Grasp the tails in both hands.
(3) Hold the dressing directly over the wound with the white side down and place it directly on the wound (Figure 3-15 A).
(4) Hold the dressing in place with one hand (the casualty may assist if able). Wrap the top tail over the top of the head and bring it down in front of the ear (on the side opposite the wound), under the chin (Figure 3-15 B) and up over the dressing to a point just above the ear (on the wound side).

Figure 3-14: Side of head or cheek wound

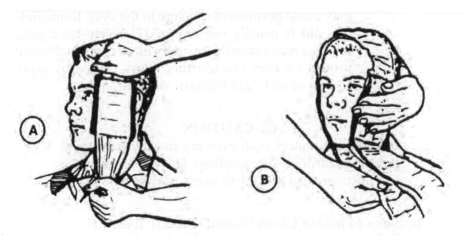

Figure 3-15: Dressing placed directly on wound. Top tail wrapped over top of head, down in front of ear, and under chin (Illustrations A and B)

> ✍ **NOTE**
> When possible, avoid covering the casualty's ear with the dressing, as this will decrease his ability to hear.

(5) Bring the second tail under the chin, up in front of the ear (on the side opposite the wound), and over the head to meet the other tail (on the wound side) (Figure 3-16).

(6) Cross the two tails (on the wound side) (Figure 3-17) and bring one end across the forehead (above the eyebrows) to

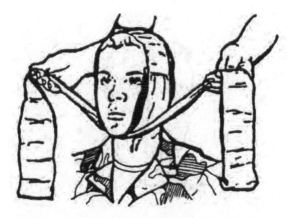

Figure 3-16: Bringing second tail under the chin

Figure 3-17: Crossing the tails on the side of the wound

Figure 3-18: Tying the tails of the dressing in a nonslip knot

a point just in front of the opposite ear (on the uninjured side).

(7) Wrap the other tail around the back of the head (at the base of the skull), and tie the two ends just in front of the ear on the uninjured side with a non slip knot (Figure 3-18).

c. Ear Injuries. Lacerated (cut) or avulsed (torn) ear tissue may not, in itself, be a serious injury. Bleeding, or the drainage of fluids from the ear canal, however, may be a sign of a head injury, such as a skull fracture. DO NOT attempt to stop the flow from the inner ear canal nor put anything into the ear canal to block it. Instead, you should cover the ear lightly with a dressing. For minor cuts or wounds to the external ear, apply a cravat bandage as follows:

(1) Place the middle of the bandage over the ear (Figure 3-19 A).

(2) Cross the ends, wrap them in opposite directions around the head, and tie them (Figures 3-19 B and 3-19 C).

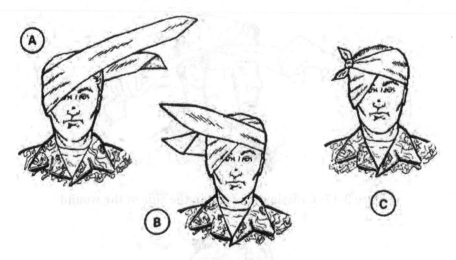

Figure 3-19: Applying cravat bandage to ear (Illustrated A thru C)

(3) If possible, place some dressing material between the back of the ear and the side of the head to avoid crushing the ear against the head with the bandage.

d. Nose Injuries. Nose injuries generally produce bleeding. The bleeding may be controlled by placing an ice pack over the nose, or pinching the nostrils together. The bleeding may also be controlled by placing torn gauze (rolled) between the upper teeth and the lip.

> ⚠ **CAUTION**
>
> DO NOT attempt to remove objects inhaled in the nose. An untrained person who/removes such an object could worsen the casualty's condition and cause permanent/injury.

e. Jaw Injuries. Before applying a bandage to a casualty's jaw, remove all foreign material from the casualty's mouth. If the casualty is unconscious, check for obstructions in the airway. When applying the bandage, allow the jaw enough freedom to permit passage of air and drainage from the mouth.

(1) Apply bandages attached to field first aid dressing to the jaw. After dressing the wound, apply the bandages

using the same technique illustrated in Figures 3-5 through 3-8.

> **⚐ NOTE**
> The dressing and bandaging procedure outlined for the jaw serves a twofold purpose In addition to stopping the bleeding and protecting the wound, it also immobilizes a fractured jaw.

(2) Apply a cravat bandage to the jaw.
    (a) Place the bandage under the chin and carry it sends upward. Adjust the bandage to make one end longer than the other (Figure 3-20 A).
    (b) Take the longer end over the top of the head to meet the short end at the temple and cross the ends over (Figure 3-20 B).
    (c) Take the ends in opposite directions to the other side of the head and tie them over the part of the bandage that was applied first (Figure 3-20 C).

> **⚐ NOTE**
> The cravat bandage technique is used to immobilize a fractured jaw or to maintain a sterile dressing that does not have tail bandages attached.

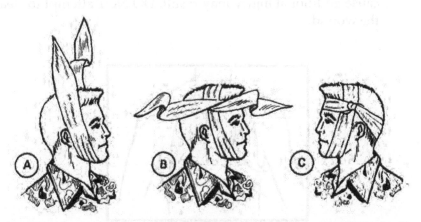

Figure 3-20: Applying cravat bandage to jaw (Illustrated A thru C).

SECTION III. GIVE PROPER FIRST AID FOR CHEST AND ABDOMINAL WOUNDS AND BURN INJURIES

3-9. Chest Wounds (081-831-1026)

Chest injuries may be caused by accidents, bullet or missile wounds, stab wounds, or falls. These injuries can be serious and may cause death quickly if proper treatment is not given. A casualty with a chest injury may complain of pain in the chest or shoulder area; he may have difficulty with his breathing. His chest may not rise normally when he breathes. The injury may cause the casualty to cough up blood and to have a rapid or a weak heartbeat. A casualty with an open chest wound has a punctured chest wall. The sucking sound heard when he breathes is caused by air leaking into his chest cavity. This particular type of wound is dangerous and will collapse the injured lung (Figure 3-21). Breathing becomes difficult for the casualty because the wound is open. The soldier's life may depend upon how quickly you make the wound airtight.

3-10. Chest Wound(s) Procedure (081-831-1026)

* a. Evaluate the Casualty (081-831-1000). Be prepared to perform lifesaving measures. The basic lifesaving measures may include clearing the airway, rescue breathing, treatment for shock, and/or bleeding control.

  b. Expose the Wound. If appropriate, cut or remove the casualty's clothing to expose the entire area of the wound. Remember, DO NOT remove clothing that is stuck to the wound because additional injury may result. DO NOT attempt to clean the wound.

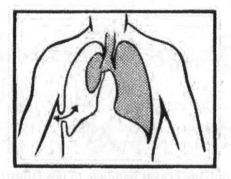

Figure 3-21: Collapsed lung

✍ NOTE

Examine the casualty to see if there is an entry and/or exit wound. If there are two wounds (entry, exit), perform the same procedure for both wounds. Treat the more serious (heavier bleeding, larger) wound first. It may be necessary to improvise a dressing for the second wound by using strips of cloth, such as a torn T-shirt, or whatever material is available. Also, listen for sucking sounds to determine if the chest wall is punctured.

⚠ CAUTION

If there is an object extending from (impaled in) the wound, DO NOT remove the object. Apply a dressing around the object and use additional improvised bulky materials/dressings (use the cleanest materials available) to buildup the area around the object. Apply a supporting bandage over the bulky materials to hold them in place.

⚠ CAUTION (081-831-1026)

DO NOT REMOVE protective clothing in a chemical environment. Apply dressings over the protective clothing.

c. Open the Casualty's Field Dressing Plastic Wrapper. The plastic wrapper is used with the field dressing to create an airtight seal. If a plastic wrapper is not available, or if an additional wound needs to be treated; cellophane, foil, the casualty's poncho, or similar material maybe used. The covering should be wide enough to extend 2 inches or more beyond the edges of the wound in all directions.

   (1) Tear open one end of the casualty's plastic wrapper covering the field dressing. Be careful not to destroy the wrapper and DO NOT touch the inside of the wrapper.

   (2) Remove the inner packet (field dressing).

   (3) Complete tearing open the empty plastic wrapper using as much of the wrapper as possible to create a flat surface.

d. Place the Wrapper Over the Wound (081-831-1026). Place the inside surface of the plastic wrapper directly over the wound when the casualty exhales and hold it in place (Figure 3-22). The casualty may hold the plastic wrapper in place if he is able.

e. Apply the Dressing to the Wound (081-831-1026).

(1) Use your free hand and shake open the field dressing (Figure 3-23).

(2) Place the white side of the dressing on the plastic wrapper covering the wound (Figure 3-24).

Figure 3-22: Open chest wound sealed with plastic wrapper

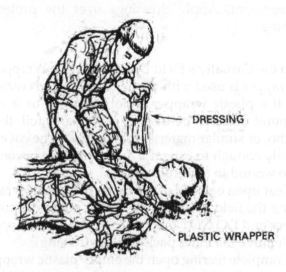

DRESSING

PLASTIC WRAPPER

Figure 3-23: Shaking open the field dressing

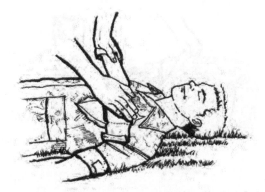

Figure 3-24: Field dressing placed on plastic wrapper

✍ **NOTE (081-831-1026)**
Use the casualty's field dressing, not your own.

---

(3) Have the casualty breathe normally.
(4) While maintaining pressure on the dressing, grasp one tail of the field dressing with the other hand and wrap it around the casualty's back.
(5) Wrap the other tail in the opposite direction, bringing both tails over the dressing (Figure 3-25).
(6) Tie the tails into a non slip knot in the center of the dressing after the casualty exhales and before he inhales. This will aid in maintaining pressure on the bandage after it has been tied (Figure 3-26). Tie the dressing firmly enough

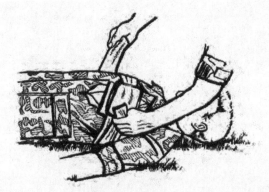

**Figure 3-25: Tails of field dressing wrapped around casualty
in opposite direction**

Figure 3-26: Tails of dressing tied into nonslip knot over center of dressing

to secure the dressing without interfering with the casualty's breathing.

✍ NOTE (081-831-1026)

When practical, apply direct manual pressure over the dressing for 5 to 10 minutes to help control the bleeding.

---

    f. Position the Casualty (081-831-1026). Position the casualty on his injured side or in a sitting position, whichever makes breathing easier (Figure 3-27).

    g. Seek Medical Aid. Contact medical personnel.

---

*WARNING

Even if an airtight dressing has been placed properly, air may still enter the chest cavity without having means to escape. This causes a life-threatening condition called tension pneumothorax. If the casualty's condition (for example, difficulty breathing, shortness of breath, restlessness, or grayness of skin in a dark-skinned individual [or blueness in an individual with light skin]) worsens after placing the dressing, quickly lift or remove, then replace the airtight dressing.

Figure 3-27: Casualty positioned (lying) on injured side

3-11. Abdominal Wounds

The most serious abdominal wound is one in which an object penetrates the abdominal wall and pierces internal organs or large blood vessels. In these instances, bleeding may be severe and death can occur rapidly.

3-12. Abdominal Wound(s) Procedure (081-831-1025)

a. Evaluate the Casualty. Be prepared to perform basic lifesaving measures. It is necessary to check for both entry and exit wounds. If there are two wounds (entry and exit), treat the wound that appears more serious first (for example, the heavier bleeding, protruding organs, larger wound, and so forth). It may be necessary to improvise dressings for the second wound by using strips of cloth, a T-shirt, or the cleanest material available.

b. Position the Casualty. Place and maintain the casualty on his back with his knees in an upright (flexed) position (Figure 3-28). The knees-up position helps relieve pain, assists in the treatment of shock, prevents further exposure of the bowel (intestines) or abdominal organs, and helps relieve abdominal pressure by allowing the abdominal muscles to relax.

c. Expose the Wound.

(1) Remove the casualty's loose clothing to expose the wound. However, DO NOT attempt to remove clothing that is stuck to the wound; it may cause further injury. Thus, remove any loose clothing from the wound but leave in place the clothing that is stuck.

> ⚠ CAUTION (081-831-1000 and 081-831-1025)
> DO NOT REMOVE protective clothing in a chemical environment. Apply dressings over the protective clothing.

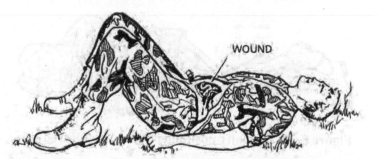

PLACE CASUALTY ON BACK TO PREVENT FURTHER EXPOSURE OF THE
BOWEL UNLESS OTHER WOUNDS PREVENT SUCH ACTION. FLEX
CASUALTY'S KNEES TO RELAX ABDOMINAL MUSCLES AND ANY INTERNAL
PRESSURE.

Figure 3-28: Casualty positioned (lying) on back with knees (flexed) up

BEFORE APPLYING DRESSINGS, CAREFULLY PLACE PROTRUDING ORGANS
NEAR THE WOUND TO PROTECT THEM AND CONTROL CONTAMINATION.

Figure 3-29: Protruding organs placed near wound

(2) Gently pick up any organs which may be on the ground.
Do this with a clean, dry dressing or with the cleanest
available material, Place the organs on top of the casualty's abdomen (Figure 3-29).

✍ **NOTE (081-831-1025)**

DO NOT probe, clean, or try to remove any foreign
object from the abdomen.
DO NOT touch with bare hands any exposed organs.
DO NOT push organs back inside the body.

---

d. Apply the Field Dressing. Use the casualty's field dressing not
your own. If the field dressing is not large enough to cover the

entire wound, the plastic wrapper from the dressing may be used to cover the wound first (placing the field dressing on top). Open the plastic wrapper carefully without touching the inner surface, if possible. If necessary other improvised dressings may be made from clothing, blankets, or the cleanest materials available because the field dressing and/or wrapper may not be large enough to cover the entire wound.

> **WARNING**
>
> If there is an object extending from the wound, DO NOT remove it. Place as much of the wrapper over the wound as possible without dislodging or moving the object. DO NOT place the wrapper over the object.

(1) Grasp the tails in both hands.

(2) Hold the dressing with the white, or cleanest, side down directly over the wound.

(3) Pull the dressing open and place it directly over the wound (Figure 3-30). If the casualty is able, he may hold the dressing in place.

(4) Hold the dressing in place with one hand and use the other hand to wrap one of the tails around the body.

IF THE DRESSING WRAPPER IS LARGE ENOUGH TO EXTEND WELL BEYOND THE PROTRUDING BOWEL. THE STERILE SIDE OF THE DRESSING WRAPPER CAN BE PLACED DIRECTLY OVER THE WOUND, WITH THE FIELD DRESSING ON THE TOP.

Figure 3-30: Dressing placed directly over the wound

(5) Wrap the other tail in the opposite direction until the dressing is completely covered. Leave enough of the tail for a knot.

(6) Loosely tie the tails with a non slip knot at the casualty's side (Figure 3-31).

> **WARNING**
>
> When dressing is applied, DO NOT put pressure on the wound or exposed internal parts, because pressure could cause further injury (vomiting, ruptured intestines, and so forth). Therefore, tie the dressing ties (tails) loosely at casualty's side, not directly over the dressing.

(7) Tie the dressing firmly enough to prevent slipping without applying pressure to the-wound-site (Figure 3-32).

Figure 3-31: Dressing applied and tails tied with a nonslip knot

Figure 3-32: Field dressing covered with improvised material and loosely tied

Field dressings can be covered with improvised reinforcement material (cravats, strips of torn T-shirt, or other cloth), if available, for additional support and protection. Tie improvised bandage on the opposite side of the dressing ties firmly enough to prevent slipping but without applying additional pressure to the wound.

> ⚠ CAUTION (081-831-1025)
> DO NOT give casualties with abdominal wounds food nor water (moistening the lips is allowed).

  e. Seek Medical Aid. Notify medical personnel.

3-13. Burn Injuries
Burns often cause extreme pain, scarring, or even death. Proper treatment will minimize further injury of the burned area. Before administering the proper first aid, you must be able to recognize the type of burn to be treated. There are four types of burns: (1) thermal burns caused by fire, hot objects, hot liquids, and gases or by nuclear blast or fire ball; (2) electrical burns caused by electrical wires, current, or lightning; (3) chemical burns caused by contact with wet or dry chemicals or white phosphorus (WP)—from marking rounds and grenades; and (4) laser burns.

3-14. First Aid for Burns (081-831-1007)

  a. Eliminate the Source of the Burn. The source of the burn must be eliminated before any evaluation or treatment of the casualty can occur.
  (1) Remove the casualty quickly and cover the thermal burn with any large non synthetic material, such as a field jacket. Roll the casualty on the ground to smother (put out) the flames (Figure 3-33).

> ⚠ CAUTION
> Synthetic materials, such as nylon, may melt and cause further injury.

  (2) Remove the electrical burn casualty from the electrical source by turning off the electrical current. DO NOT attempt to turn off the electricity if the source is not close

Figure 3-33: Casualty covered and rolled on ground

by. Speed is critical, so DO NOT waste unnecessary time. If the electricity cannot be turned off, wrap any nonconductive material (dry rope, dry clothing, dry wood, and so forth) around the casualty's back and shoulders and drag the casualty away from the electrical source (Figure 3-34). DO NOT make body-to-body contact with the casualty or touch any wires because you could also become an electrical burn casualty.

> **WARNING**
> High voltage electrical burns may cause temporary unconsciousness, difficulties in breathing, or difficulties with the heart (heartbeat).

(3) Remove the chemical from the burned casualty. Remove liquid chemicals by flushing with as much water as possible. If water is not available, use any nonflammable fluid to flush chemicals off the casualty. Remove dry chemicals by brushing off loose particles (DO NOT use the bare surface of your hand because you could become a chemical burn casualty) and then flush with large amounts of water, if available. If large amounts of water are not available, then NO water should be applied because small amounts of water applied to a dry chemical burn may cause

**Figure 3-34: Casualty removed from electrical source
(using nonconductive material)**

a chemical reaction. When white phosphorous strikes the skin, smother with water, a wet cloth, or wet mud. Keep white phosphorous covered with a wet material to exclude air which will prevent the particles from burning.

> **WARNING**
>
> Small amounts of water applied to a dry chemical burn may cause a chemical reaction, transforming the dry chemical into an active burning substance.

(4) Remove the laser burn casualty from the source. (NOTE: Lasers produce a narrow amplified beam of light. The word laser means Light Amplification by Stimulated Emission of Radiation and sources include range finders, weapons/guidance, communication systems, and weapons simulations such as MILES.) When removing the casualty from the laser beam source, be careful not to enter the beam or you may become a casualty. Never look directly at the beam source and if possible, wear appropriate eye protection.

✍ NOTE

After the casualty is removed from the source of the burn, he should be evaluated for conditions requiring basic lifesaving measures (Evaluate the Casualty).

---

b. Expose the Burn. Cut and gently lift away any clothing covering the burned area, without pulling clothing over the burns. Leave in place any clothing that is stuck to the burns. If the casualty's hands or wrists have been burned, remove jewelry if possible without causing further injury (rings, watches, and so forth) and place in his pockets. This prevents the necessity to cut off jewelry since swelling usually occurs as a result of a burn.

⚠ CAUTION (081-831-1007)
- DO NOT lift or cut away clothing if in a chemical environment. Apply the dressing directly over the casualty's protective clothing.
- DO NOT attempt to decontaminate skin where blisters have formed.

---

c. Apply a Field Dressing to the Burn.
   (1) Grasp the tails of the casualty's dressing in both hands.
   (2) Hold the dressing directly over the wound with the white (sterile) side down, pull the dressing open, and place it directly over the wound. If the casualty is able, he may hold the dressing in place.
   (3) Hold the dressing in place with one hand and use the other hand to wrap one of the tails around the limbs or the body.
   (4) Wrap the other tail in the opposite direction until the dressing is completely covered.
   (5) Tie the tails into a knot over the outer edge of the dressing. The dressing should be applied lightly over the burn. Ensure that dressing is applied firmly enough to prevent it from slipping.

✍ NOTE

Use the cleanest improvised dressing material available if a field dressing is not available or if it is not large enough for the entire wound.

---

d. Take the Following Precautions (081-831-1007):
   - DO NOT place the dressing over the face or genital area.
   - DO NOT break the blisters.
   - DO NOT apply grease or ointments to the burns.
     For electrical burns, check for both an entry and exit burn from the passage of electricity through the body. Exit burns may appear on any area of the body despite location of entry burn.
   - For burns caused by wet or dry chemicals, flush the burns with large amounts of water and cover with a dry dressing.
   - For burns caused by white phosphorus (WP), flush the area with water, then cover with a wet material, dressing, or mud to exclude the air and keep the WP particles from burning.
   - For laser burns, apply a field dressing.
   - If the casualty is conscious and not nauseated, give him small amounts of water.

e. Seek Medical Aid. Notify medical personnel.

SECTION IV. APPLY PROPER BANDAGES TO UPPER AND LOWER EXTREMITIES

3-15. Shoulder Bandage

a. To apply bandages attached to the field first aid dressing–
   (1) Take one bandage across the chest and the other across the back and under the arm opposite the injured shoulder.
   (2) Tie the ends with a non slip knot (Figure 3-35).

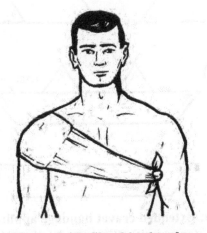

Figure 3-35: Shoulder bandage

b. To apply a cravat bandage to the shoulder or armpit–

   (1) Make an extended cravat bandage by using two triangular bandages (Figure 3-36 A); place the end of the first triangular bandage along the base of the second one (Figure 3-36 B).

   (2) Fold the two bandages into a single extended bandage (Figure 3-36 C).

   (3) Fold the extended bandage into a single cravat bandage (Figure 3-36 D). After folding, secure the thicker part (overlap) with two or more safety pins (Figure 3-36 E).

   (4) Place the middle of the cravat bandage under the armpit so that the front end is longer than the back end and safety pins are on the outside (Figure 3-36 F).

   (5) Cross the ends on top of the shoulder (Figure 3-36 G).

   (6) Take one end across the back and under the arm on the opposite side and the other end across the chest. Tie the ends (Figure 3-36 H).

Be sure to place sufficient wadding in the armpit. DO NOT tie the cravat bandage too tightly. Avoid compressing the major blood vessels in the armpit.

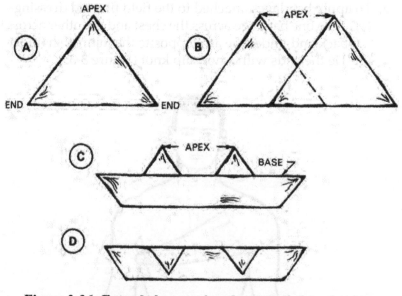

Figure 3-36: Extended cravat bandage applied to shoulder (or armpit) (Illustrated A thru H)

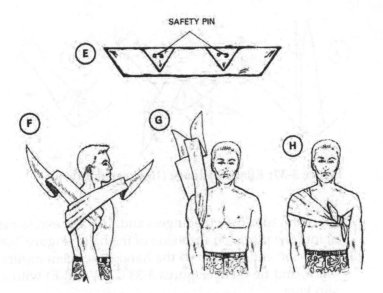

Figure 3-36: *(Continued)*

3-16. Elbow Bandage

To apply a cravat bandage to the elbow–

   a. Bend the arm at the elbow and place the middle of the cravat at the point of the elbow bringing the ends upward (Figure 3-37 A).

   b. Bring the ends across, extending both downward (Figure 3-37 B).

   c. Take both ends around the arm and tie them with a non slip knot at the front of the elbow (Figure 3-37 C).

> ### ⚠ CAUTION
> If an elbow fracture is suspected, DO NOT bend the elbow; bandage it in an extended position.

3-17. Hand Bandage

   a. To apply a triangular bandage to the hand–

      (1) Place the hand in the middle of the triangular bandage with the wrist at the base of the bandage (Figure 3-38 A). Ensure that the fingers are separated with absorbent material to prevent chafing and irritation of the skin.

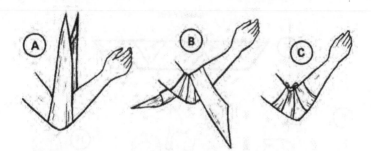

Figure 3-37: Elbow bandage (Illustrated A thru C)

    (2) Place the apex over the fingers and tuck any excess materi-
al into the pleats on each side of the hand (Figure 3-38 B).

    (3) Cross the ends on top of the hand, take them around the
wrist, and tie them (Figures 3-38 C, D, and E) with a non
slip knot.

  b. To apply a cravat bandage to the palm of the hand–

    (1) Lay the middle of the cravat over the palm of the hand
with the ends hanging down on each side (Figure 3-39 A).

    (2) Take the end of the cravat at the little finger across the back
of the hand, extending it upward over the base of the thumb;
then bring it downward across the palm (Figure 3-39 B).

    (3) Take the thumb end across the back of the hand, over the
palm, and through the hollow between the thumb and
palm (Figure3-39 C).

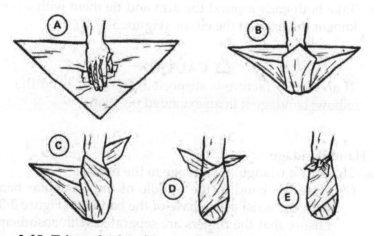

Figure 3-38: Triangular bandage applied to hand (Illustrated A thru E)

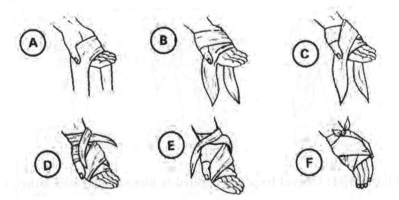

**Figure 3-39: Cravat bandage applied to palm of hand
(Illustrated A thru F)**

(4) Take the ends to the back of the hand and cross them; then bring them up over the wrist and cross them again (Figure 3-39 D).

(5) Bring both ends down and tie them with a non slip knot on top of the wrist (Figure 3-39 E and F).

3-18. Leg (Upper and Lower) Bandage

To apply a cravat bandage to the leg–

a. Place the center of the cravat over the dressing (Figure 3-40 A).

b. Take one end around and up the leg in a spiral motion and the other end around and down the leg in a spiral motion, overlapping part of each preceding turn (Figure 3-40 B).

c. Bring both ends together and tie them (Figure 3-40 C) with a non slip knot.

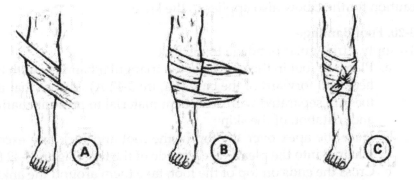

Figure 3-40: Cravat bandage applied to leg (Illustrated A thru C)

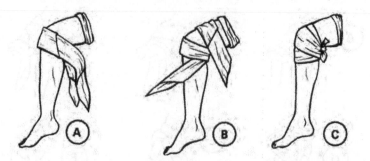

Figure 3-41: Cravat bandage applied to knee (Illustrated A thru C)

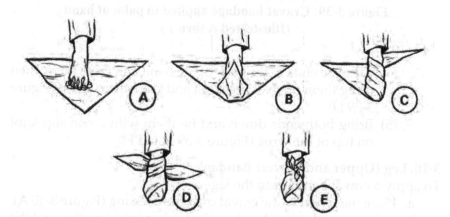

Figure 3-42: Triangular bandage applied to foot (Illustrated A thru E)

3-19. Knee Bandage
To apply a cravat bandage to the knee as illustrated in Figure 3-41, use the same technique applied in bandaging the elbow. The same caution for the elbow also applies to the knee.

3-20. Foot Bandage
To apply a triangular bandage to the foot–
   a. Place the foot in the middle of the triangular bandage with the heel well forward of the base (Figure 3-42 A). Ensure that the toes are separated with absorbent material to prevent chafing and irritation of the skin.
   b. Place the apex over the top of the foot and tuck any excess material into the pleats on each side of the foot (Figure 3-42 B).
   c. Cross the ends on top of the foot, take them around the ankle, and tie them at the front of the ankle (Figure 3-42 C, D, and E).

CHAPTER 4

FIRST AID FOR FRACTURES

A fracture is any break in the continuity of a bone. Fractures can cause total disability or in some cases death. On the other hand, they can most often be treated so there is complete recovery. A great deal depends upon the first aid the individual receives before he is moved. First aid includes immobilizing the fractured part in addition to applying lifesaving measures. The basic splinting principle is to immobilize the joints above and below any fracture.

4-1. Kinds of Fractures
See figure 4-1 for detailed illustration.

    a. Closed Fracture. A closed fracture is a broken bone that does not break the overlying skin. Tissue beneath the skin may be damaged. A dislocation is when a joint, such as a knee, ankle, or shoulder, is not in proper position. A sprain is when the

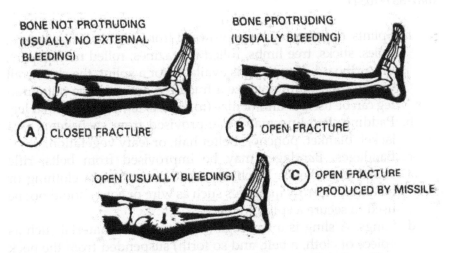

BONE NOT PROTRUDING (USUALLY NO EXTERNAL BLEEDING)

BONE PROTRUDING (USUALLY BLEEDING)

(A) CLOSED FRACTURE

(B) OPEN FRACTURE

OPEN (USUALLY BLEEDING)

(C) OPEN FRACTURE PRODUCED BY MISSILE

Figure 4-1: Kinds of fractures (Illustrated A thru C)

connecting tissues of the joints have been torn. Dislocations and sprains should be treated as closed fractures.

b. Open Fracture. An open fracture is a broken bone that breaks (pierces) the overlying skin. The broken bone may come through the skin, or a missile such as a bullet or shell fragment may go through the flesh and break the bone. An open fracture is contaminated and subject to infection.

4-2. Signs/Symptoms of Fractures (081-831-1000)
Indications of a fracture are deformity, tenderness, swelling, pain, inability to move the injured part, protruding bone, bleeding, or discolored skin at the injury site. A sharp pain when the individual attempts to move the part is also a sign of a fracture. DO NOT encourage the casualty to move the injured part in order to identify a fracture since such movement could cause further damage to surrounding tissues and promote shock. If you are not sure whether a bone is fractured, treat the injury as a fracture.

4-3. Purposes of Immobilizing Fractures
A fracture is immobilized to prevent the sharp edges of the bone from moving and cutting tissue, muscle, blood vessels, and nerves. This reduces pain and helps prevent or control shock. In a closed fracture, immobilization keeps bone fragments from causing an open wound and prevents contamination and possible infection. Splint to immobilize.

4-4. Splints, Padding, Bandages, Slings, and Swathes (081-831-1034)

a. Splints. Splints may be improvised from such items as boards, poles, sticks, tree limbs, rolled magazines, rolled newspapers, or cardboard. If nothing is available for a splint, the chest wall can be used to immobilize a fractured arm and the uninjured leg can be used to immobilize (to some extent) the fractured leg.

b. Padding. Padding may be improvised from such items as a jacket, blanket, poncho, shelter half, or leafy vegetation.

c. Bandages. Bandages may be improvised from belts, rifle slings, bandoleers, kerchiefs, or strips torn from clothing or blankets. Narrow materials such as wire or cord should not be used to secure a splint in place.

d. Slings. A sling is a bandage (or improvised material such as apiece of cloth, a belt, and so forth) suspended from the neck to support an upper extremity. Also, slings may be improvised by using the tail of a coat or shirt, and pieces torn from such items as clothing and blankets. The triangular bandage is ideal for this purpose. Remember that the casualty's hand should be higher than his elbow, and the sling should be applied so that the supporting pressure is on the uninjured side.

e. Swathes. Swathes are any bands (pieces of cloth, pistol belts, and so forth) that are used to further immobilize a splinted fracture. Triangular and cravat bandages are often used as or referred to as swathe bandages. The purpose of the swathe is to immobilize, therefore, the swathe bandage is placed above and/or below the fracture—not over it.

4-5. Procedures for Splinting Suspected Fractures (081-831-1034)
Before beginning first aid treatment for a fracture, gather whatever splinting materials are available. Materials may consist of splints, such as wooden boards, branches, or poles. Other splinting materials include padding, improvised cravats, and/or bandages, Ensure that splints are long enough to immobilize the joint above and below the suspected fracture. If possible, use at least four ties (two above and two below the fracture) to secure the splints. The ties should be non slip knots and should be tied away from the body on the splint.

*a. Evaluate the Casualty (081-831-1000). Be prepared to perform my necessary lifesaving measures. Monitor the casualty for development of conditions which may require you to perform necessary basic lifesaving measures. These measures include clearing the airway, rescue breathing, preventing shock, and/or bleeding control.

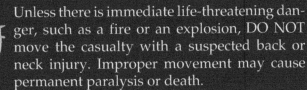

WARNING (081-831-1000)
Unless there is immediate life-threatening danger, such as a fire or an explosion, DO NOT move the casualty with a suspected back or neck injury. Improper movement may cause permanent paralysis or death.

WARNING (081-831-1000)
In a chemical environment, DO NOT remove any protective clothing. Apply the dressing/splint over the clothing.

b. Locate the Site of the Suspected Fracture. Ask the casualty for the location of the injury. Does he have any pain? Where is it tender? Can he move the extremity? Look for an unnatural position of the extremity. Look for a bone sticking out (protruding).

c. Prepare the Casualty for Splinting the Suspected Fracture (081-831-1034).

 (1) Reassure the casualty. Tell him that you will be taking care of him and that medical aid is on the way.

 (2) Loosen any tight or binding clothing.

 (3) Remove all the jewelry from the casualty and place it in the casualty's pocket. Tell the casualty you are doing this because if the jewelry is not removed at this time and swelling occurs later, further bodily injury can occur.

✍ NOTE

Boots should not be removed from the casualty unless they are needed to stabilize a neck injury, or there is actual bleeding from the foot.

---

d. Gather Splinting Materials (081-831-1034). If standard splinting materials (splints, padding, cravats, and so forth) are not available, gather improvised materials. Splints can be improvised from wooden boards, tree branches, poles, rolled newspapers or magazines. Splints should be long enough to reach beyond the joints above and below the suspected fracture site. Improvised padding, such as a jacket, blanket, poncho, shelter half, or leafy vegetation may be used. A cravat can be improvised from a piece of cloth, a large bandage, a shirt, or a towel. Also, to immobilize a suspected fracture of an arm or a leg, parts of the casualty's body may be used. For example, the chest wall may be used to immobilize an arm; and the uninjured leg may be used to immobilize the injured leg.

✍ NOTE

If splinting material is not available and suspected fracture CANNOT be splinted, then swathes, or a combination of swathes and slings can be used to immobilize an extremity.

---

e. Pad the Splints (081-831-1034). Pad the splints where they touch any bony part of the body, such as the elbow, wrist, knee, ankle, crotch, or armpit area. Padding prevents excessive pressure to the area.

f. Check the Circulation Below the Site of the Injury (081-831-1034).

(1) Note any pale, white, or bluish-gray color of the skin which may indicate impaired circulation. Circulation can also be checked by depressing the toe/fingernail beds and observing how quickly the color returns. A slower return of pink color to the injured side when compared with the uninjured side indicates a problem with circulation. Depressing the toe/fingernail beds is a method to use to check the circulation in a dark-skinned casualty.

(2) Check the temperature of the injured extremity. Use your hand to compare the temperature of the injured side with the uninjured side of the body. The body area below the injury maybe colder to the touch indicating poor circulation.

(3) Question the casualty about the presence of numbness, tightness, cold, or tingling sensations.

WARNING

Casualties with fractures to the extremities may show impaired circulation, such as numbness, tingling, cold and/or pale to blue skin. These casualties should be evacuated by medical personnel and treated as soon as possible. Prompt medical treatment may prevent possible loss of the limb.

WARNING

If it is an open fracture (skin is broken; bone(s) may be sticking out), DO NOT ATTEMPT TO PUSH BONE(S) BACK UNDER THE SKIN. Apply a field dressing to protect the area. See Task 081-831-1016, Put on a Field or Pressure Dressing.

g. Apply the Splint in Place (081-831-1034).

(1) Splint the fracture(s) in the position found. DO NOT attempt to reposition or straighten the injury. If it is an open fracture, stop the bleeding and protect the wound. (See Chapter 2, Section II, for detailed information.) Cover all wounds with field dressings before applying a splint. Remember to use the casualty's field dressing, not your own. If bones are protruding (sticking out), DO NOT attempt to push them back under the skin. Apply dressings to protect the area.

(2) Place one splint on each side of the arm or leg. Make sure that the splints reach, if possible, beyond the joints above and below the fracture.

(3) Tie the splints. Secure each splint in place above and below the fracture site with improvised (or actual) cravats. Improvised cravats, such as strips of cloth, belts, or whatever else you have, may be used. With minimal motion to the injured areas, place and tie the splints with the bandages. Push cravats through and under the natural body curvatures (spaces), and then gently position improvised cravats and tie in place. Use non slip knots. Tie all knots on the splint away from the casualty (Figure 4-2). DO NOT tie cravats directly over suspected fracture/dislocation site.

h. Check the Splint for Tightness (081-831-1034).

(1) Check to be sure that bandages are tight enough to securely hold splinting materials in place, but not so tight that circulation is impaired.

(2) Recheck the circulation after application of the splint. Check the skin color and temperature. This is to ensure that the bandages holding the splint in place have not been tied too tightly. A finger tip check can be made by inserting the tip of the finger between the wrapped tails and the skin.

Figure 4-2: Nonslip knots tied away from casualty

(3) Make any adjustment without allowing the splint to become ineffective.

i. Apply a Sling if Applicable (081-831-1034). An improvised sling may be made from any available non stretching piece of cloth, such as a fatigue shirt or trouser, poncho, or shelter half. Slings may also be improvised using the tail of a coat, belt, or a piece of cloth from a blanket or some clothing. See Figure 4-3 for an illustration of a shirt tail used for support. A pistol belt or trouser belt also may be used for support (Figure 4-4). A sling should place the supporting pressure on

Figure 4-3: Shirt tail used for support

Figure 4-4: Belt used for support

the casualty's uninjured side. The supported arm should have the hand positioned slightly higher than the elbow.

(1) Insert the splinted arm in the center of the sling (Figure 4-5).

(2) Bring the ends of the sling up and tie them at the side (or hollow) of the neck on the uninjured side (Figure 4-6).

(3) Twist and tuck the corner of the sling at the elbow (Figure 4-7).

j. Apply a swathe if Applicable (081-831-1034). You may use any large piece of cloth, such as a soldier's belt or pistol belt, to improvise a swathe. A swathe is any band (a piece of cloth) or wrapping used to further immobilize a fracture. When splints are unavailable, swathes, or a combination of swathes and slings can be used to immobilize an extremity.

Figure 4-5: Arm inserted in center of improvised sling

Figure 4-6: Ends of improvised sling tied to side of neck

Figure 4-7: Corner of sling twisted and tucked at elbow

Figure 4-8: Arm immobilized with strip of clothing

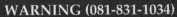

WARNING (081-831-1034)
The swathe should not be placed directly on top of the injury, but positioned either above and/or below the fracture site.

(1) Apply swathes to the injured arm by wrapping the swathe over the injured arm, around the casualty's back and under the arm on the uninjured side. Tie the ends on the uninjured side (Figure 4-8).

(2) A swathe is applied to an injured leg by wrapping the swathe(s) around both legs and securing it on the uninjured side.

k. Seek Medical Aid. Notify medical personnel, watch closely for development of life-threatening conditions, and if necessary, continue to evaluate the casualty.

4-6. Upper Extremity Fractures (081-831-1034)

Figures 4-9 through 4-16 show how to apply slings, splints, and cravats (swathes) to immobilize and support fractures of the upper extremities. Although the padding is not visible in some of the illustrations, it is always preferable to apply padding along the injured part for the length of the splint and especially where it touches any bony parts of the body.

METHOD 1

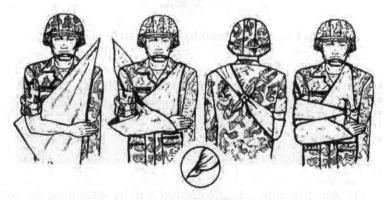

METHOD 2

Figure 4-9: Application of triangular bandage to form sling (two methods) (Illustrated A and B)

Figure 4-10: Completing sling sequence by twisting and tucking the corner of the sling at the elbow

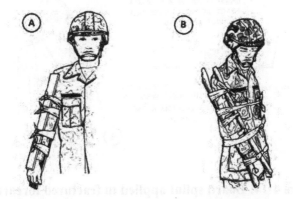

Figure 4-11: Board splints applied to fractured elbow when elbow is not bent (two methods) (081-831-1034) (Illustrated A and B)

Figure 4-12: Chest wall used as splint for upper arm fracture when no splint is available (Illustrated A and B)

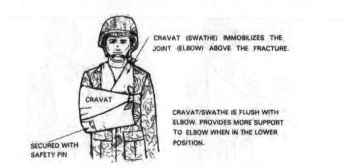

Figure 4-13: Chest wall, sling, and cravat used to immobilize fractured elbow when elbow is bent

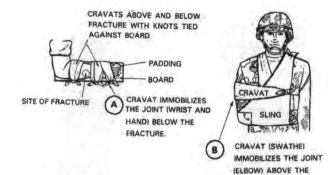

Figure 4-14: Board splint applied to fractured forearm (Illustrated A and B)

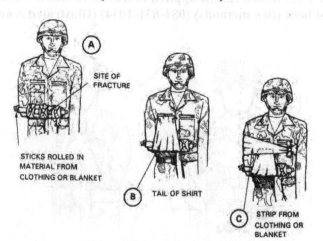

Figure 4-15: Fractured forearm or wrist splinted with sticks and supported with tail of shirt and strips of material (Illustrated A thru C)

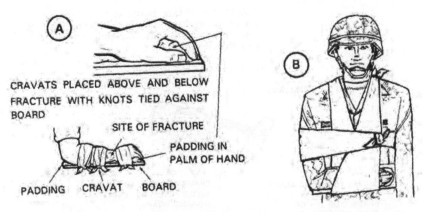

**Figure 4-16: Board splint applied to fractured wrist and hand
(Illustrated A thru C)**

4-7. Lower Extremity Fractures (081-831-1034)

Figures 4-17 through 4-22 show how to apply splints to immobilize fractures of the lower extremities. Although padding is not visible in some of the figures, it is preferable to apply padding along the injured part for the length of the splint and especially where it touches any bony parts of the body.

4-8. Jaw, Collarbone, and Shoulder Fractures

    a. Apply a cravat to immobilize a fractured jaw as illustrated in Figure 4-23. Direct all bandaging support to the top of the casualty's head, not to the back of his neck. If incorrectly placed,

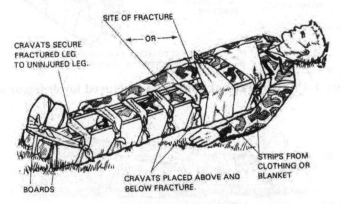

**Figure 4-17: Board splint applied to fractured hip or thigh
(081-831-1034)**

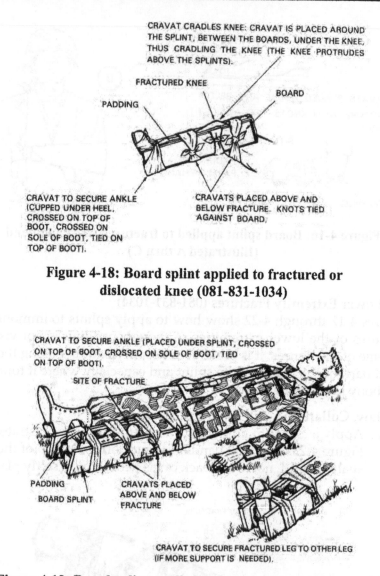

Figure 4-18: Board splint applied to fractured or dislocated knee (081-831-1034)

Figure 4-19: Board splint applied to fractured lower leg or ankle

Figure 4-20: Improvised splint applied to fractured lower leg or ankle

the bandage will pull the casualty's jaw back and interfere with his breathing.

> ⚠ **CAUTION**
>
> Casualties with lower jaw (mandible) fractures cannot be laid flat on their backs because facial muscles will relax and may cause an airway obstruction.

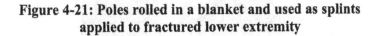

Figure 4-21: Poles rolled in a blanket and used as splints applied to fractured lower extremity

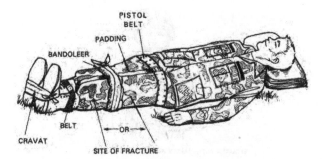

Figure 4-22: Uninjured leg used as splint for fractured leg (anatomical splint)

b. Apply two belts, a sling, and a cravat to immobilize a fractured collarbone, as illustrated in Figure 4-24.

c. Apply a sling and a cravat to immobilize a fractured or dislocated shoulder, using the technique illustrated in Figure 4-25.

4-9. Spinal Column Fractures (081-831-1000)

It is often impossible to be sure a casualty has a fractured spinal column. Be suspicious of any back injury, especially if the casualty has fallen or if his back has been sharply struck or bent. If a casualty has

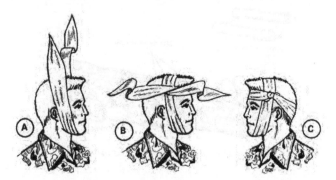

Figure 4-23: Fractured jaw immobilized (Illustrated A thru C)

Figure 4-24: Application of belts, sling, and cravat to immobilize a collarbone

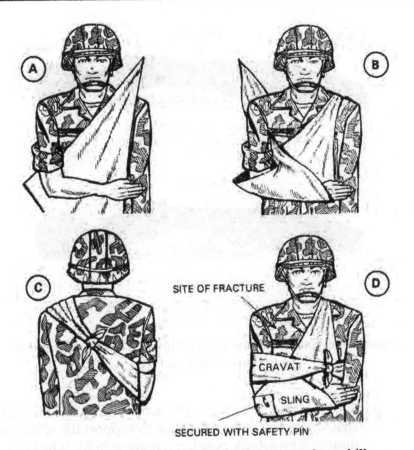

SITE OF FRACTURE

CRAVAT

SLING

SECURED WITH SAFETY PIN

Figure 4-25: Application of sling and cravat to immobilize a fractured or dislocated shoulder (Illustrated A thru D)

received such an injury and does not have feeling in his legs or cannot move them, you can be reasonably sure that he has a severe back injury which should be treated as a fracture. Remember, if the spine is fractured, bending it can cause the sharp bone fragments to bruise or cut the spinal cord and result in permanent paralysis (Figure 4-26A). The spinal column must maintain a swayback position to remove pressure from the spinal cord.

   a. If the Casualty Is Not to Be Transported (081-831-1000) Until Medical Personnel Arrive—

     • Caution him not to move. Ask him if he is in pain or if he is unable to move any part of his body.

     • Leave him in the position in which he is found. DO NOT move any part of his body.

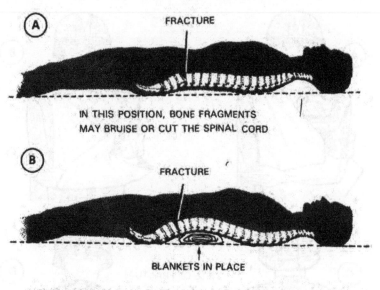

Figure 4-26: Spinal column must maintain a swayback position (Illustrated A and B)

- Slip a blanket, if he is lying face up, or material of similar size, under the arch of his back to support the spinal column in a swayback position (Figure 4-26 B). If he is lying face down, DO NOT put anything under any part of his body.
b. If the Casualty Must Be Transported to A Safe Location Before Medical Personnel Arrive—
   - And if the casualty is in a face-up position, transport him by litter or use a firm substitute, such as a wide board or a flat door longer than his height. Loosely tie the casualty's wrists together over his waistline, using a cravat or a strip of cloth. Tie his feet together to prevent the accidental dropping or shifting of his legs. Lay a folded blanket across the litter where the arch of his back is to be placed. Using a four-man team (Figure 4-27), place the casualty on the litter without bending his spinal column or his neck.
     - The number two, three, and four men position themselves on one side of the casualty; all kneel on one knee along the side of the casualty. The number one man

WRISTS TIED LOOSELY

FEET TIED LOOSELY

Figure 4-27: Placing face-up casualty with fractured back onto litter

positions himself to the opposite side of the casualty. The number two, three, and four men gently place their hands under the casualty. The number one man on the opposite side places his hands under the injured part to assist.

o When all four men are in position to lift, the number two man commands, "PREPARE TO LIFT" and then, "LIFT." All men, in unison, gently lift the casualty about 8 inches. Once the casualty is lifted, the number one man recovers and slides the litter under the casualty, ensuring that the blanket is in proper position. The number one man then returns to his original lift position (Figure 4-27).

       ○ When the number two man commands, "LOWER CASUALTY," all men, in unison, gently lower the casualty onto the litter.
- And if the casualty is in a face-down position, he must be transported in this same position. The four-man team lifts him onto a regular or improvised litter, keeping the spinal column in a swayback position. If a regular litter is used, first place a folded blanket on the litter at the point where the chest will be placed.

4-10. Neck Fractures (081-831-1000)
A fractured neck is extremely dangerous. Bone fragments may bruise or cut the spinal cord just as they might in a fractured back.

    a. **If the Casualty Is Not to Be Transported (081-831-1000) Until Medical Personnel Arrive—**
- Caution him not to move. Moving may cause death.
- Leave the casualty in the position in which he is found. If his neck/head is in an abnormal position, immediately immobilize the neck/head. Use the procedure stated below.
  - ○ Keep the casualty's head still, if he is lying face up, raise his shoulders slightly, and slip a roll of cloth that has the bulk of a bath towel under his neck (Figure 4-28). The roll should be thick enough to arch his neck only slightly, leaving the back of his head on the ground. DO NOT bend his neck or head forward. DO NOT raise or twist his head. Immobilize the casualty's head (Figure 4-29). Do this by padding heavy objects such as rocks or his boots and placing them on each side of his head. If it is necessary to use boots, first fill them with stones, gravel, sand, or dirt and tie them tightly at the top. If necessary, stuff pieces of material in the top of the boots to secure the contents.
  - ○ DO NOT move the casualty if he is lying face down. Immobilize the head/neck by padding heavy objects and placing them on each side of his head. DO NOT put a roll of cloth under the neck. DO NOT bend the neck or head, nor roll the casualty onto his back.

    b. **If the Casualty Must be Prepared for Transportation Before Medical Personnel Arrive—**

Figure 4-28: Casualty with roll of cloth (bulk) under neck

Figure 4-29: Immobilization of fractured neck

- And he has a fractured neck, at least two persons are needed because the casualty's head and trunk must be moved in unison.
- The two persons must work in close coordination (Figure 4-30) to avoid bending the neck.
- Place a wide board lengthwise beside the casualty. It should extend at least 4 inches beyond the casualty's head and feet (Figure 4-30 A).
- If the casualty is lying face up, the number one man steadies the casualty's head and neck between his hands. At the same time the number two man positions one foot and one knee against the board to prevent it from slipping, grasps the casualty underneath his shoulder and hip, and gently slides him onto the board (Figure 4-30 B).
- If the casualty is lying face down, the number one man steadies the casualty's head and neck between his hands,

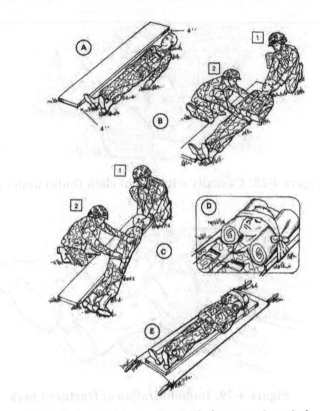

Figure 4-30: Preparing casualty with fractured neck for transportation (Illustrated A thru E)

while the number two man gently rolls the casualty over onto the board (Figure 4-30 C).

- The number one man continues to steady the casualty's head and neck. The number two man simultaneously raises the casualty's shoulders slightly, places padding under his neck, and immobilizes the casualty's head (Figures 4-30 D, and E). The head may be immobilized with the casualty's boots, with stones rolled in pieces of blanket, or with other material.

- Secure any improvised supports in position with a cravat or strip of cloth extended across the casualty's forehead and under the board (Figure 4-30 D).

- Lift the board onto a litter or blanket in order to transport the casualty (Figure 4-30 E).

CHAPTER 5

FIRST AID FOR CLIMATIC INJURIES

It is desirable, but not always possible, for an individual's body to become adjusted (acclimatized) to an environment. Physical condition determines the time adjustment, and trying to rush it is ineffective. Even those individuals in good physical condition need time before working or training in extremes of hot or cold weather. Climate-related injuries are usually preventable; prevention is both an individual and leadership responsibility. Several factors contribute to health and well-being in any environment: diet, sleep/rest, exercise, and suitable clothing. These factors are particularly important in extremes of weather. Diet, especially, should be suited to an individual's needs in a particular climate. A special diet undertaken for any purpose should be done so with appropriate supervision. This will ensure that the individual is getting a properly balanced diet suited to both climate and personal needs, whether for weight reduction or other purposes. The wearing of specialized protective gear or clothing will sometimes add to the problem of adjusting to a particular climate. Therefore, soldiers should exercise caution and judgment in adding or removing specialized protective gear or clothing.

5-1. Heat Injuries (081-831-1008)
Heat injuries are environmental injuries that may result when a soldier is exposed to extreme heat, such as from the sun or from high temperatures. Prevention depends on availability and consumption of adequate amounts of water. Prevention also depends on proper clothing and appropriate activity levels. Acclimatization and protection from undue heat exposure are also very important. Identification of high risk personnel (basic trainees, troops with previous history of heat injury, and overweight soldiers) helps both the leadership and the individual prevent and cope with climatic conditions. Instruction on living and working ingot climates also contributes toward prevention.

141

> ✍ **NOTE**
> Salt tablets should not be used in the prevention of heat injury. Usually, eating field rations or liberal salting of the garrison diet will provide enough salt to replace what is lost through sweating in hot weather.

a. Diet. A balanced diet usually provides enough salt even in hot weather. But when people are on reducing or other diets, salt may need to come from other sources. DO NOT use salt tablets to supplement a diet. Anyone on a special diet (for whatever purpose) should obtain professional help to work out a properly balanced diet.

b. Clothing.

(1) The type and amount of clothing and equipment a soldier wears and the way he wears it also affect the body and its adjustment to the environment. Clothing protects the body from radiant heat. However, excessive or tight-fitting clothing, web equipment, and packs reduce ventilation needed to cool the body. During halts, rest stops, and other periods when such items are not needed, they should be removed, mission permitting.

(2) The individual protective equipment (IPE) protects the soldier from chemical and biological agents. The equipment provides a barrier between him and a toxic environment. However, a serious problem associated with the chemical overgarment is heat stress. The body normally maintains a heat balance, but when the overgarment is worn the body sometimes does not function properly. Overheating may occur rapidly. Therefore, strict adherence to mission oriented protective posture (MOPP) levels directed by your commander is important. This will keep those heat related injuries caused by wearing the IPE to a minimum. See FM 3-4 for further information on MOPP.

c. Prevention. The ideal fluid replacement is water. The availability of sufficient water during work or training in hot weather is very important. The body, which depends on water to help cool itself, can lose more than a quart of water per hour through sweat. Lost fluids must be replaced quickly. Therefore, during these work or training periods, you should drink at least one canteen full of water every hour. In extremely hot climates or extreme temperatures, drink at least a full canteen of water every half hour, if possible. In such hot climates, the body depends mainly upon sweating to keep it cool, and water intake must be maintained to allow sweating to continue. Also, keep in mind that a person who has suffered one heat injury is likely to suffer another. Before a heat injury casualty returns to work, he should have recovered well enough not to risk a recurrence. Other conditions which may increase heat stress and cause heat injury include infections, fever, recent illness or injury, overweight, dehydration, exertion, fatigue, heavy meals, and alcohol. In all this, note that salt tablets should not be used as a preventive measure.

d. Categories. Heat injury can be divided into three categories: heat cramps, heat exhaustion, and heatstroke.

e. First Aid. Recognize and give first aid for heat injuries.

WARNING
Casualty should be continually monitored for development of conditions which may require the performance of necessary basic lifesaving measures, such as: clearing the airway, performing mouth-to-mouth resuscitation, preventing shock, and/or bleeding control.

⚠ * CAUTION
DO NOT use salt solution in first aid procedures for heat injuries.

(1) Check the casualty for signs and symptoms of heat cramps (081-831-1008).
  • Signs/Symptoms. Heat cramps are caused by an imbalance of chemicals (called electrolytes) in the body as a result of excessive sweating. This condition causes the casualty to exhibit:
    ○ Muscle cramps in the extremities (arms and legs).
    ○ Muscle cramps of the abdomen.

- o Heavy (excessive) sweating (wet skin).
- o Thirst.
- Treatment.
  - o Move the casualty to a cool or shady area (or improvise shade).
  - o Loosen his clothing (if not in a chemical environment).
  - o Have him slowly drink at least one canteen full of cool water.
  - o Seek medical aid should cramps continue.

WARNING

DO NOT loosen the casualty's clothing if in a chemical environment. 160-065 0-94-3

(2) Check the casualty for signs and symptoms of heat exhaustion (081-831-1008).
  - Signs/Symptoms which occur often. Heat exhaustion is caused by loss of water through sweating without adequate fluid replacement. It can occur in an otherwise fit individual who is involved in tremendous physical exertion in any hot environment. The signs and symptoms are similar to those which develop when a person goes into a state of shock.
    - o Heavy (excessive) sweating with pale, moist, cool skin.
    - o Headache.
    - o Weakness.
    - o Dizziness.
    - o Loss of appetite.
  - Signs/Symptoms which occur sometimes.
    - o Heat cramps.
    - o Nausea—with or without vomiting.
    - o Urge to defecate.
    - o Chills (gooseflesh).
    - o Rapid breathing.
    - o Tingling of hands and/or feet.
    - o Confusion.
  - Treatment.
    - o Move the casualty to a cool or shady area (or improvise shade).

      o Loosen or remove his clothing and boots (unless in a chemical environment). Pour water on him and fan him (unless in a chemical environment).

      o Have him slowly drink at least one canteen full of cool water. Elevate his legs.

      o If possible, the casualty should not participate in strenuous activity for the remainder of the day.

      o Monitor the casualty until the symptoms are gone, or medical aid arrives.

(3) Check the casualty for signs and symptoms of heatstroke (sometimes called "sunstroke") (081-831-1008).

> **WARNING**
>
> Heatstroke must be considered a medical emergency which may result in death if treatment is delayed.

- Signs/Symptoms. A casualty suffering from prolonged time. It is caused by failure of the body's cooling mechanisms. Inadequate sweating is a factor. The casualty's skin is red (flushed), hot, and dry. He may experience weakness, dizziness, confusion, headaches, seizures, nausea (stomach pains), and his respiration and pulse may be rapid and weak. Unconsciousness and collapse may occur suddenly.

- Treatment. Cool casualty immediately by—
  - o Moving him to a cool or shaded area (or improvise shade).
  - o Loosening or removing his clothing (except in a chemical environment).
  - o *Spraying or pouring water on him; fanning him to permit a coolant effect of evaporation.
  - o Massaging his extremities and skin which increases the blood flow to those body areas, thus aiding the cooling process.
  - o Elevating his legs.
  - o Having him slowly drink at least one canteen full of water if he is conscious.

> ✍ **NOTE**
> Start cooling casualty immediately. Continue cooling while awaiting transportation and during the evacuation.

- Medical aid. Seek medical aid because the casualty should be transported to a medical treatment facility as soon as possible. Do not interrupt cooling process or lifesaving measures to seek help.
- Casualty should be continually monitored for development of conditions which may require the performance of necessary basic lifesaving measures, such as clearing the airway, mouth-to-mouth resuscitation, preventing shock, and/or bleeding control.

f. Table. See Table 5-1 for further information.

Table 5-1: Sun or Heat Injuries (081-831-1008)

INJURIES	SIGNS/SYMPTOMS	FIRST AID*
Heat cramps	The casualty experiences muscle cramps of arms, legs, and/or stomach. The casualty may also have heavy sweating (wet skin) and extreme thirst.	1. Move the casualty to a shady area or improvise shade and loosen his clothing.+ 2. Give him large amounts of cool water slowly. 3. Monitor the casualty and give him more water as tolerated. 4. Seek medical aid if the cramps continue.
Heat exhaustion	The casualty *often* experiences profuse (heavy) sweating with pale, moist, cool skin; headache, weakness, dizziness, and/or loss of appetite.	1. Move the casualty to a cool, shady area or improvise shade and loosen/remove his clothing.+ 2. Pour water on him and fan him to permit coolant effect of evaporation. 3. Have him slowly drink at least one canteen full of water.

(continued)

Table 5-1: *(Continued)*

Heat exhaustion *Continued.*	The casualty *sometimes* experiences heat cramps, nausea (with or without vomiting), urge to defecate, chills (gooseflesh), rapid breathing, confusion, and tingling of the hands and/or feet.	4. Elevate the casualty's legs. 5. Seek medical aid if symptoms continue; monitor the casualty until the symptoms are gone or medical aid arrives.
Heatstroke[#] (sunstroke)	The casualty stops sweating (red [flushed] hot, dry skin). He first may experience headache, dizziness, nausea, fast pulse and respiration, seizures, and mental confusion. He may collapse and suddenly become unconscious. *THIS IS A MEDICAL EMERGENCY.*	1. Move the casualty to a cool, shady area or improvise shade and loosen or remove his clothing, remove the outer garments and protective clothing if the situation permits.[+] ★2. Start cooling the casualty immediately. Spray or pour water on him. Fan him. Massage his extremities and skin. 3. Elevate his legs. 4. If conscious, have him slowly drink at least one canteen full of water. 5. SEEK MEDICAL AID. CONTINUE COOLING WHILE AWAITING TRANSPORT AND DURING EVACUATION. EVACUATE AS SOON AS POSSIBLE. PERFORM ANY NECESSARY LIFESAVING MEASURES.

*The *first aid procedure* for heat related injuries caused by wearing *individual protective equipment* is to move the casualty to a clean area and give him water to drink.

+When in a chemical environment, DO NOT loosen/remove the casualty's clothing.

#Can be fatal if not treated promptly and correctly.

5-2. Cold Injuries (081-831-1009)
Cold injuries are most likely to occur when an unprepared individual is exposed to winter temperatures. They can occur even with proper planning and equipment. The cold weather and the type of combat operation in which the individual is involved impact on whether he is likely to be injured and to what extent. His clothing, his physical condition, and his mental makeup also are determining factors. However, cold injuries can usually be prevented. Well-disciplined and well-trained individuals can be protected even in the most adverse circumstances. They and their leaders must know the hazards of exposure to the cold. They must know the importance of personal hygiene, exercise, care of the feet and hands, and the use of protective clothing.

a. Contributing Factors
   (1) Weather. Temperature, humidity, precipitation, and wind modify the loss of body heat. Low temperatures and low relative humidity-dry cold—promote frostbite. Higher temperatures, together with moisture, promote immersion syndrome. Wind chill accelerates the loss of body heat and may aggravate cold injuries. These principles and risks apply equally to both men and women.
   (2) Type of combat operation. Defense, delaying, observation-post, and sentinel duties do create to a greater extent—fear, fatigue, dehydration, and lack of nutrition. These factors further increase the soldier's vulnerability to cold injury. Also, a soldier is more likely to receive a cold injury if he is—
      • Often in contact with the ground.
      • Immobile for long periods, such as while riding in a crowded vehicle.
      • Standing in water, such as in a foxhole.
      • Out in the cold for days without being warmed.
      • Deprived of an adequate diet and rest.
      • Not able to take care of his personal hygiene.
   (3) Clothing. The soldier should wear several layers of loose clothing. He should dress as lightly as possible consistent with the weather to reduce the danger of excessive perspiration and subsequent chilling. It is better for the body to be slightly cold and generating heat than excessively warm and sweltering toward dehydration. He should remove a layer or two of clothing before doing any hard work. He should replace the clothing when work is completed. Most

cold injuries result from soldiers having too few clothes available when the weather suddenly turns colder. Wet gloves, shoes, socks, or any other wet clothing add to the cold injury process.

⚠ CAUTION
In a chemical environment DO NOT take off protective chemical gear.

(4) Physical makeup. Physical fatigue contributes to apathy, which leads to inactivity, personal neglect, carelessness, and reduced heat production. In turn, these increase the risk of cold injury. Soldiers with prior cold injuries have a higher-than-normal risk of subsequent cold injury, not necessarily involving the part previously injured.

(5) Psychological factor. Mental fatigue and fear reduces the body's ability to rewarm itself and thus increases the incidence of cold injury. The feelings of isolation imposed by the environment are also stressful. Depressed and/or unresponsive soldiers are also vulnerable because they are less active. These soldiers tend to be careless about precautionary measures, especially warming activities, when cold injury is a threat.

b. Signs/Symptoms. Once a soldier becomes familiar with the factors that contribute to cold injury, he must learn to recognize cold injury signs/symptoms.

(1) Many soldiers suffer cold injury without realizing what is happening to them. They may be cold and generally uncomfortable. These soldiers often do not notice the injured part because it is already numb from the cold.

(2) Superficial cold injury usually can be detected by numbness, tingling, or "pins and needles" sensations. These signs/symptoms often can be relieved simply by loosening boots or other clothing and by exercising to improve circulation. In more serious cases involving deep cold injury, the soldier often is not aware that there is a problem until the affected part feels like a stump or block of wood.

(3) Outward signs of cold injury include discoloration of the skin at the site of injury. In light-skinned persons, the skin first reddens and then becomes pale or waxy white.

In dark-skinned persons, grayness in the skin is usually evident. An injured foot or hand feels cold to the touch. Swelling may be an indication of deep injury. Also note that blisters may occur after rewarming the affected parts. Soldiers should work in pairs—buddy teams—to check each other for signs of discoloration and other symptoms. Leaders should also be alert for signs of cold injuries.

c. Treatment Considerations. First aid for cold injuries depends on whether they are superficial or deep. Cases of superficial cold injury can be adequately treated by warming the affected part using body heat. For example, this can be done by covering cheeks with hands, putting fingertips under armpits, or placing feet under the clothing of a buddy next to his belly. The injured part should NOT be massaged, exposed to a fire or stove, rubbed with snow, slapped, chafed, or soaked in cold water. Walking on injured feet should be avoided. Deep cold injury (frostbite) is very serious and requires more aggressive first aid to avoid or to minimize the loss of parts of the fingers, toes, hands, or feet. The sequence for treating cold injuries depends on whether the condition is life-threatening. That is, PRIORITY is given to removing the casualty from the cold. Other-than-cold injuries are treated either simultaneously while waiting for evacuation to a medical treatment facility or while reroute to the facility.

✍ NOTE

The injured soldier should be evacuated at once to a place where the affected part can be rewarmed under medical supervision.

---

d. Conditions Caused by Cold. Conditions caused by cold are chilblain, immersion syndrome (immersion foot/trench foot), frostbite, snow blindness, dehydration, and hypothermia.

(1) Chilblain.
- Signs/Symptoms. Chilblain is caused by repeated prolonged exposure of bare skin at temperatures from 60°F, to 32°F, or 200°F for acclimated, dry, unwashed skin. The area may be acutely swollen, red, tender, and hot with itchy skin. There may be no loss of skin tissue in untreated cases but continued exposure may lead to infected, ulcerated, or bleeding lesions.

- Treatment. Within minutes, the area usually responds to locally applied body heat. Rewarm the affected part by applying firm steady pressure with your hands, or placing the affected part under your arms or against the stomach of a buddy. DO NOT rub or massage affected areas. Medical personnel should evaluate the injury, because signs and symptoms of tissue damage may be slow to appear.
- Prevention. Prevention of chilblain depends on basic cold injury prevention methods. Caring for and wearing the uniform properly and staying dry (as far as conditions permit) are of immediate importance.

(2) Immersion syndrome (immersion foot/trench foot). Immersion foot and trench foot are injuries that result from fairly long exposure of the feet to wet conditions at temperatures from approximately 50°F to 32°F. Inactive feet in damp or wet socks and boots, or tightly laced boots which impair circulation are even more susceptible to injury. This injury can be very serious; it can lead to loss of toes or parts of the feet. If exposure of the feet has been prolonged and severe, the feet may swell so much that pressure closes the blood vessels and cuts off circulation. Should an immersion injury occur, dry the feet thoroughly; and evacuate the casualty to a medical treatment facility by the fastest means possible.

- Signs/Symptoms. At first, the parts of the affectedfoot are cold and painless, the pulse is weak, and numbness may be present. Second, the parts may feel hot, and burning and shooting pains may begin. In later stages, the skin is pale with a bluish cast and the pulse decreases. Other signs/symptoms that may follow are blistering, swelling, redness, heat, hemorrhages (bleeding), and gangrene.
- Treatment. Treatment is required for all stages of immersion syndrome injury. Rewarm the injured part gradually by exposing it to warm air. DO NOT massage it. DO NOT moisten the skin and DO NOT apply heat or ice. Protect it from trauma and secondary infections. Dry, loose clothing or several layers of warm coverings are preferable to extreme heat. Under no circumstances should the injured part be exposed

to an open fire. Elevate the injured part to relieve the swelling. Evacuate the casualty to a medical treatment facility as soon as possible. When the part is rewarmed, the casualty often feels a burning sensation and pain. Symptoms may persist for days or weeks even after rewarming.

- Prevention. Immersion syndrome can be prevented by good hygienic care of the feet and avoiding moist conditions for prolonged periods. Changing socks at least daily (depending on environmental conditions) is also a preventive measure. Wet socks can be air dried, then can be placed inside the shirt to warm them prior to putting them on.

(3) Frostbite. Frostbite is the injury of tissue caused from exposure to cold, usually below 32°F depending on the wind chill factor, duration of exposure, and adequacy of protection. Individuals with a history of cold injury are likely to be more easily affected for an indefinite period. The body parts most easily frostbitten are the cheeks, nose, ears, chin, forehead, wrists, hands, and feet. Proper treatment and management depend upon accurate diagnosis. Frostbite may involve only the skin (superficial), or it may extend to a depth below the skin (deep). Deep frostbite is very serious and requires more aggressive first aid to avoid or to minimize the loss of parts of the fingers, toes, hands, or feet.

WARNING
Casualty should be continually monitored for development of conditions which may require the performance of necessary basic lifesaving measures, such as clearing the airway, performing mouth-to-mouth resuscitation, preventing shock, and/or bleeding control.

- Progressive signs/symptoms (081-831-1009).
  ○ Loss of sensation, or numb feeling in any part of the body.
  ○ Sudden blanching (whitening) of the skin of the affected part, followed by a momentary "tingling" sensation.

- o   Redness of skin in light-skinned soldiers;
- o   grayish coloring in dark-skinned individuals.
- o   Blister.
- o   Swelling or tender areas.
- o   Loss of previous sensation of pain in affected area.
- o   Pale, yellowish, waxy-looking skin.
- o   Frozen tissue that feels solid (or wooden) to the touch.

⚠ CAUTION

Deep frostbite is a very serious injury and requires immediate first aid and subsequent medical treatment to avoid or minimize loss of body parts.

- • Treatment (081-831-1009).
  - o   Face, ears, and nose. Cover the casualty's affected area with his and/or your bare hands until sensation and color return.
  - o   Hands. Open the casualty's field jacket and shirt. (In a chemical environment never remove the clothing.) Place the affected hands under the casualty's armpits. Close the field jacket and shirt to prevent additional exposure.
  - o   Feet. Remove the casualty's boots and socks if he does not need to walk any further to receive additional treatment. (Thawing the casualty's feet and forcing him to walk on them will cause additional pain/injury.) Place the affected feet under clothing and against the body of another soldier.

WARNING (081-831-1009)

DO NOT attempt to thaw the casualty's feet or other seriously frozen areas if he will be required to walk or travel to receive further treatment. The casualty should avoid walking, if possible, because there is less danger in walking while the feet are frozen than after they have been thawed. Thawing in the field increases the possibilities of infection, gangrene, or other injury.

✍ NOTE

Thawing may occur spontaneously during transportation to the medical facility; this cannot be avoided since the body in general must be kept warm.

In all of the above areas, ensure that the casualty is kept warm and that he is covered (to avoid further injury). Seek medical treatment as soon as possible. Reassure the casualty, protect the affected area from further injury by covering it lightly with a blanket or any dry clothing, and seek shelter out of the wind. Remove/minimize constricting clothing and increase insulation. Ensure that the casualty exercises as much as possible, avoiding trauma to the injured part, and is prepared for pain when thawing occurs. Protect the frostbitten part from additional injury. DO NOT rub the injured part with snow or apply cold water soaks. DO NOT warm the part by massage or exposure to open fire because the frozen part may be burned due to the lack of feeling. DO NOT use ointments or other medications. DO NOT manipulate the part in any way to increase circulation. DO NOT allow the casualty to use alcohol or tobacco because this reduces the body's resistance to cold. Remember, when freezing extends to a depth below the skin, it involves a much more serious injury. Extra care is required to reduce or avoid the chances of losing all or part of the toes or feet. This also applies to the fingers and hands.

- Prevention. Prevention of frostbite or any cold injury depends on adequate nutrition, hot meals, and warm fluids. Other cold injury preventive factors are proper clothing and maintenance of general body temperature. Fatigue, dehydration, tobacco, and alcoholic beverages should be avoided.
  - Sufficient clothing must be worn for protection against cold and wind. Layers of clothing that can be removed and replaced as needed are the most effective. Every effort must be made to keep clothing and body as dry as possible. This includes avoiding any excessive perspiration by removing and replacing layers of clothing. Socks should be changed whenever the feet become moist or wet. Clothing and equipment should be properly fitted to avoid any interference with blood circulation. Improper blood circulation reduces the amount

of heat that reaches the extremities. Tight fitting socks, shoes, and hand wear are especially hazardous in very cold climates. The face needs extra protection against high winds, and the ears need massaging from time to time to maintain circulation. Hands may be used to massage and warm the face. By using the buddy system, individuals can watch each other's face for signs of frostbite to detect it early and keep tissue damage to a minimum. A mask or headgear tunneled in front of the face guards against direct wind injury. Fingers and toes should be exercised to keep them warm and to detect any numbness. Wearing windproof leather gloves or mittens and avoiding kerosene, gasoline, or alcohol on the skin are also preventive measures. Cold metal should not be touched with bare skin; doing so could result in severe skin damage.

- o Adequate clothing and shelter are also necessary during periods of inactivity.

(4) Snow blindness. Snow blindness is the effect that glare from an ice field or snowfield has on the eyes. It is more likely to occur in hazy, cloudy weather than when the sun is shining. Glare from the sun will cause an individual to instinctively protect his eyes. However, in cloudy weather, he may be overconfident and expose his eyes longer than when the threat is more obvious. He may also neglect precautions such as the use of protective eyewear. Waiting until discomfort (pain) is felt before using protective eyewear is dangerous because a deep burn of the eyes may already have occurred.

- • Signs/Symptoms. Symptoms of snow blindness area sensation of grit in the eyes with pain in and over the eyes, made worse by eyeball movement. Other signs/symptoms are watering, redness, headache, and increased pain on exposure to light. The same condition that causes snow blindness can cause snow burn of skin, lips, and eyelids. If a snow burn is neglected, the result is the same as a sunburn.

- • Treatment. First aid measures consist of blindfolding or covering the eyes with a dark cloth which stops painful eye movement. Complete rest is desirable.

If further exposure to light is not preventable, the eyes should be protected with dark bandages or the darkest glasses available. Once unprotected exposure to sunlight stops, the condition usually heals in a few days without permanent damage. The casualty should be evacuated to the nearest medical facility.

- Prevention. Putting on protective eye wear is essential not only to prevent injury, but to prevent further injury if any has occurred. When protective eye wear is not available, an emergency pair can be made from a piece of wood or cardboard cut and shaped to the width of the face. Cut slits for the eyes and attach strings to hold the improvised glasses in place. Slits are made at the point of vision to allow just enough space to see and reduce the risk of injury. Blackening the eyelids and face around the eyes absorbs some of the harmful rays.

(5) Dehydration. Dehydration occurs when the body loses too much fluid, salt, and minerals. A certain amount of body fluid is lost through normal body processes. A normal daily intake of food and liquids replaces these losses. When individuals are engaged in any strenuous exercises or activities, an excessive amount of fluid and salt is lost through sweat. This excessive loss creates an imbalance of fluids, and dehydration occurs when fluid and salt are not replaced. It is very important to know that it can be prevented if troops are instructed in its causes, symptoms, and preventive measures. The danger of dehydration is as prevalent in cold regions as it is in hot regions. In hot weather the individual is aware of his body losing fluids and salt. He can see, taste, and feel the sweat as it runs down his face, gets into his eyes, and on his lips and tongue, and drips from his body. In cold weather, however, it is extremely difficult to realize that this condition exists. The danger of dehydration in cold weather operations is a serious problem. In cold climates, sweat evaporates so rapidly or is absorbed so thoroughly by layers of heavy clothing that it is rarely visible on the skin. Dehydration also occurs during cold weather operations because drinking is inconvenient. Dehydration will weaken or incapacitate

a casualty for a few hours, or sometimes, several days. Because rest is an important part of the recovery process, casualties must take care that limited movement during their recuperative period does not enhance the risk of becoming a cold weather casualty.

- Signs/Symptoms. The symptoms of cold weather dehydration are similar to those encountered in heat exhaustion. The mouth, tongue, and throat become parched and dry, and swallowing becomes difficult. The casualty may have nausea with or without vomiting along with extreme dizziness and fainting. The casualty may also feel generally tired and weak and may experience muscle cramps (especially in the legs). Focusing eyes may also become difficult.
- Treatment. The casualty should be kept warm and his clothes should be loosened to allow proper circulation. Shelter from wind and cold will aid in this treatment. Fluid replacement, rest, and prompt medical treatment are critical. Medical personnel will determine the need for salt replacement.
- Prevention. These general preventive measures apply for both hot and cold weather. Sufficient additional liquids should be consumed to offset excessive body losses of these elements. The amount should vary according to the individual and the type of work he is doing (light, heavy, or very strenuous). Rest is equally important as a preventive measure. Each individual must realize that any work that must be done while bundled in several layers of clothing is extremely exhausting. This is especially true of any movement by foot, regardless of the distance.

(6) Hypothermia (general cooling). In intense cold a soldier may become both mentally and physically numb, thus neglecting essential tasks or requiring more time and effort to achieve them. Under some conditions (particularly cold water immersion), even a soldier in excellent physical condition may die in a matter of minutes. The destructive influence of cold on the body is called hypothermia. This means bodies lose heat faster than they can produce it. Frostbite may occur without hypothermia

when extremities do not receive sufficient heat from central body stores. The reason for this is inadequate circulation and/or inadequate insulation. Nonetheless, hypothermia and frostbite may occur at the same time with exposure to below-freezing temperatures. An example of this is an avalanche accident. Hypothermia may occur from exposure to temperatures above freezing, especially from immersion in cold water, wet-cold conditions, or from the effect of wind. Physical exhaustion and insufficient food intake may also increase the risk of hypothermia. Excessive use of alcohol leading to unconsciousness in a cold environment can also result in hypothermia. General cooling of the entire body to a temperature below 95°F is caused by continued exposure to low or rapidly dropping temperatures, cold moisture, snow, or ice. Fatigue, poor physical condition, dehydration, faulty blood circulation, alcohol or other drug intoxication, trauma, and immersion can cause hypothermia. Remember, cold affects the body systems slowly and almost without notice. Soldiers exposed to low temperatures for extended periods may suffer ill effects even if they are well protected by clothing.

- Signs/Symptoms. As the body cools, there are several stages of progressive discomfort and impairment. A sign/symptom that is noticed immediately is shivering. Shivering is an attempt by the body to generate heat. The pulse is faint or very difficult to detect. People with temperatures around 90°F may be drowsy and mentally slow. Their ability to move may be hampered, stiff, and uncoordinated, but they may be able to function minimally. Their speech may be slurred. As the body temperature drops further, shock becomes evident as the person's eyes assume a glassy state, breathing becomes slow and shallow, and the pulse becomes weaker or absent. The person becomes very stiff and uncoordinated. Unconsciousness may follow quickly. As the body temperature drops even lower, the extremities freeze, and a deep (or core) body temperature (below 85°F) increases the risk of irregular heart action. This irregular heart action or heart standstill can result in sudden death.

- Treatment. Except in cases of the most severe hypothermia (marked by coma or unconsciousness, a weak pulse, and a body temperature of approximately 90°F or below), the treatment for hypothermia is directed towards rewarming the body evenly and without delay. Provide heat by using a hot water bottle, electric blanket, campfire, or another soldier's body heat. Always call or send for help as soon as possible and protect the casualty immediately with dry clothing or a sleeping bag. Then, move him to a warm place. Evaluate other injuries and treat them. Treatment can be given while the casualty is waiting evacuation or while he is en route. In the case of an accidental breakthrough into ice water, or other hypothermic accident, strip the casualty of wet clothing immediately and bundle him into a sleeping bag. Mouth-to-mouth resuscitation should be started at once if the casualty's breathing has stopped or is irregular or shallow. Warm liquids may be given gradually but must not be forced on an unconscious or semiconscious person because he may choke. The casualty should be transported on a litter because the exertion of walking may aggravate circulation problems. A physician should immediately treat any hypothermia casualty. Hypothermia is life-threatening until normal body temperature has been restored. The treatment of a casualty with severe hypothermia is based upon the following principles: stabilize the temperature, attempt to avoid further heat loss, handle the casualty gently, and evacuate as soon as possible to the nearest medical treatment facility! Rewarming a severely hypothermic casualty is extremely dangerous in the field due to the great possibility of such complications as rewarming shock and disturbances in the rhythm of the heartbeat.

⚠ *CAUTION

Hypothermia is a MEDICAL EMERGENCY! Prompt medical treatment is necessary. Casualties with hypothermic complications should be transported to a medical treatment facility immediately.

INJURIES	SIGNS/SYMPTOMS	FIRST AID
Chilblain	Red, swollen, hot, tender, itching skin. Continued exposure may lead to infected (ulcerated or bleeding) skin lesions.	1. Area usually responds to locally applied rewarming (body heat). 2. DO NOT rub or massage area. 3. Seek medical treatment.
Immersion foot/ Trench foot	Affected parts are cold, numb, and painless. Parts may then be hot, with burning and shooting pains. Advanced stage: skin pale with bluish cast; pulse decreases; blistering, swelling, heat, hemorrhages, and gangrene may follow.	1. Gradual rewarming by exposure to warm air. 2. DO NOT massage or moisten skin. 3. Protect affected parts from trauma. 4. Dry feet thoroughly, avoid walking. 5. Seek medical treatment.
Frostbite	Loss of sensation, or numb feeling in any part of the body. Sudden blanching (whitening) of the skin of the affected part, followed by a momentary "tingling" sensation. Redness of skin in light-skinned soldiers; grayish coloring in dark-skinned individuals. Blisters. Swelling or tender areas. Loss of previous sensation of pain in affected area. Pale,	1. Warm the area at the first sign of frostbite, using firm, steady pressure of hand, underarm or abdomen. 2. Face, ears, nose—cover area with hands (casualty's own or buddy's). 3. Hand(s)—open field jacket and place casualty's hand(s) against body, then close jacket to prevent heat loss. 4. Feet—casualty's boots/socks removed and exposed feet placed under clothing and against body of another soldier. 5. *Warning:* Do not attempt to thaw the casualty's feet or other seriously frozen areas if he will be required

Frostbite *Continued.*	yellowish, waxy-looking skin. Frozen tissue that feels solid (or wooden) to the touch.	to walk or travel to a medical center in order to receive additional treatment. The possibility of injury from walking is less when the feet are frozen than after they have been thawed. (However, if possible, avoid walking.) Thawing in the field increases the possibility of infection, gangrene, or injury. 6. Loosen or remove constricting clothing and remove any jewelry. 7. Increase insulation (cover with blanket or other dry material). Ensure casualty exercises as much as possible, avoiding trauma to injured part.
Snow Blindness	Eyes may feel scratchy. Watering, redness, headache, and increased pain with exposure to light can occur.	1. Cover the eyes with a dark cloth. 2. Seek medical treatment.
Dehydration	Similar to heat exhaustion. See Table 5-1.	1. Keep warm, loosen clothes. 2. Casualty needs fluid replacement, rest, and prompt medical treatment.
Hypothermia	Casualty is cold. Shivering stops. Core temperature is low. Consciousness may be altered. Uncoordinated movements may occur. Shock and coma may result as	*Mild Hypothermia* 1. Rewarm body evenly and without delay. (Need to provide heat source; casualty's body unable to generate heat). 2. Keep dry, protect from elements.

INJURIES	SIGNS/SYMPTOMS	FIRST AID
Hypothermia *Continued.*	body temperature drops.	3. Warm liquids may be given gradually (to conscious casualties only). ★ 4. Seek medical treatment immediately! *Severe Hypothermia* 1. Stabilize the temperature. 2. Attempt to avoid further heat loss. 3. Handle the casualty gently. 4. Evacuate to the nearest medical treatment facility as soon as possible.

★**CAUTION:** Hypothermia is a *MEDICAL EMERGENCY!* Prompt medical treatment is necessary.

⚠ **CAUTION**

The casualty is unable to generate his own body heat. Therefore, merely placing him in a blanket or sleeping bag is not sufficient.

- Prevention. Prevention of hypothermia consists of all actions that will avoid rapid and uncontrollable loss of body heat. Individuals should be properly equipped and properly dressed (as appropriate for conditions and exposure). Proper diet, sufficient rest, and general principles apply. Ice thickness must be tested before river or lake crossings. Anyone departing a fixed base by aircraft, ground vehicle, or foot must carry sufficient protective clothing and food reserves to survive during unexpected weather changes or other unforeseen emergencies. Traveling alone is never safe. Expected itinerary and arrival time should be left with responsible parties before any departure in severe weather. Anyone living in cold regions should learn how to build expedient shelters from available materials including snow.

e. Table. See Table 5-2 for further information.

CHAPTER 6

FIRST AID FOR BITES AND STINGS

Snakebites, insect bites, or stings can cause intense pain and/or swelling. If not treated promptly and correctly, they can cause serious illness or death. The severity of a snakebite depends upon: whether the snake is poisonous or nonpoisonous, the type of snake, the location of the bite, and the amount of venom injected. Bites from humans and other animals, such as dogs, cats, bats, raccoons, and rats can cause severe bruises and infection, and tears or lacerations of tissue. Awareness of the potential sources of injuries can reduce or prevent them from occurring. Knowledge and prompt application of first aid measures can lessen the severity of injuries from bites and stings and keep the soldier from becoming a serious casualty.

6-1. Types of Snakes

a. Nonpoisonous Snakes. There are approximately 130 different varieties of nonpoisonous snakes in the United States. They have oval-shaped heads and round eyes. Unlike poisonous snakes, discussed below, nonpoisonous snakes do not have fangs with which to inject venom. See Figure 6-1 for characteristics of a nonpoisonous snake.

b. Poisonous Snakes. Poisonous snakes are found throughout the world, primarily in tropical to moderate climates. Within the United States, there are four kinds: rattlesnakes, copperheads, water moccasins (cottonmouth), and coral snakes. Poisonous snakes in other parts of the world include sea snakes, the fer-de-lance, the bushmaster, and the tropical rattlesnake in tropical Central America; the Malayan pit viper in the tropical Far East; the cobra in Africa and Asia; the mamba (or black mamba) in Central and Southern Africa; and the krait in India and Southeast Asia. See Figure 6-2 for characteristics of a poisonous pit viper.

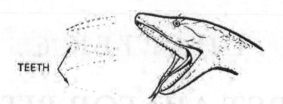

Figure 6-1: Characteristics of nonpoisonous snake

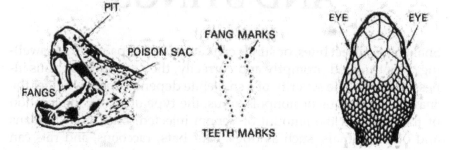

Figure 6-2: Characteristics of poisonous pit viper

   c. Pit Vipers (Poisonous). See Figure 6-3 for illustrations.
   (1) Rattlesnakes, bushmasters, copperheads, fer-de-lance, Malayan pit vipers, and water moccasins (cottonmouth) are called pit vipers because of the small, deep pits between the nostrils and eyes on each side of the head (Figure 6-2). In addition to their long, hollow fangs, these snakes have other identifying features: thick bodies, slit-like pupils of the eyes, and flat, almost triangular-shaped heads. Color markings and other identifying characteristics, such as rattles or a noticeable white interior of the mouth (cottonmouth), also help distinguish these poisonous snakes. Further identification is provided by examining the bite pattern of the wound for signs of fang entry. Occasionally there will be only one fang mark, as in the case of a bite on a finger or toe where there is no room for both fangs, or when the snake has broken off a fang.
   (2) The casualty's condition provides the best information about the seriousness of the situation, or how much time has passed since the bite occurred. Pit viper bites are characterized by severe burning pain. Discoloration and swelling around the fang marks usually begins within 5 to 10

minutes after the bite. If only minimal swelling occurs within 30 minutes, the bite will almost certainly have been from a nonpoisonous snake or possibly from a poisonous snake which did notinject venom. The venom destroys blood cells, causing a general discoloration of the skin. This reaction is followed by blisters and numbness in the affected area. Other signs which can occur are weakness, rapid pulse, nausea, shortness of breath, vomiting, and shock.

d. Corals, Cobras, Kraits, and Mambas. Corals, cobra, kraits, and mambas all belong to the same group even though they are found indifferent parts of the world. All four inject their venom through short, grooved fangs, leaving a characteristic bite pattern. See Figure 6-4 for illustration of a cobra snake.

(1) The small coral snake, found in the Southeastern United States, is brightly colored with bands of red, yellow (or almost white), and black completely encircling the body (Figure 6-5). Other nonpoisonous snakes have the same coloring, but on the coral snake found in the United States, the red ring always touches the yellow ring. To know the

TROPICAL RATTLESNAKE

MALAYAN PIT VIPER

BUSHMASTER   FER-DE-LANCE

Figure 6-3: Poisonous snakes

Figure 6-4: Cobra snake

Figure 6-5: Coral snake

difference between a harmless snake and the coral snake found in the United States, remember the following "Red on yellow will kill a fellow. Red on black, venom will lack."

(2) The venom of corals, cobras, kraits, and mambas produces symptoms different from those of pit vipers. Because there is only minimal pain and swelling, many people believe that the bite is not serious. Delayed reactions in the nervous system normally occur between 1 to 7 hours after the bite. Symptoms include blurred vision, drooping eyelids, slurred speech, drowsiness, and increased salivation and sweating. Nausea, vomiting, shock, respiratory difficulty, paralysis, convulsions, and coma will usually develop if the bite is not treated promptly.

e. Sea Snakes. Sea snakes (Figure 6-6) are found in the warm water areas of the Pacific and Indian oceans, along the coasts, and at the mouths of some larger rivers. Their venom is VERY poisonous, but their fangs are only 1/4 inch long. The first aid outlined for land snakes also applies to sea snakes.

6-2. Snakebites

If a soldier should accidentally step on or otherwise disturb a snake, it will attempt to strike. Chances of this happening while traveling along trails or waterways are remote if a soldier is alert and careful. Poisonous snakes DO NOT always inject venom when they bite or strike a person. However, all snakes may carry tetanus (lockjaw); anyone bitten by a snake, whether poisonous or nonpoisonous, should immediately seek medical attention. Poison is injected from the venom sacs through grooved or hollow fangs. Depending on the species, these fangs are either long or short. Pit vipers have long hollow fangs. These fangs are folded against the roof of the mouth and extend when the snake strikes. This allows them to strike quickly and then withdraw. Cobras, coral snakes, kraits, mambas, and sea snakes have short, grooved fangs. These snakes are less effective in their attempts to bite, since they must chew after striking to inject enough venom (poison) to be effective. See Figure 6-7 for characteristics of a poisonous snakebite. In the event you are bitten, attempt to identify and/or kill the snake. Take it to medical personnel for

Figure 6-6: Sea snake

inspection/identification. This provides valuable information to medical personnel who deal with snakebites. TREAT ALL SNAKE-BITES AS POISONOUS.

a. Venoms. The venoms of different snakes cause different effects. Pit viper venoms (hemotoxins) destroy tissue and blood cells. Cobras, adders, and coral snakes inject powerful venoms (neurotoxins) which affect the central nervous system, causing respiratory paralysis. Water moccasins and sea snakes have venom that is both hemotoxic and neurotoxic.

b. Identification. The identification of poisonous snakes is very important since medical treatment will be different for each type of venom. Unless it can be positively identified the snake should be killed and saved. When this is not possible or when doing so is a serious threat to others, identification may sometimes be difficult since many venomous snakes resemble harmless varieties. When dealing with snakebite problems in foreign countries, seek advice, professional or otherwise, which may help identify species in the particular area of operations.

*c. First Aid. Get the casualty to a medical treatment facility as soon as possible and with minimum movement. Until evacuation or treatment is possible, have the casualty lie quietly and not move anymore than necessary. The casualty should not smoke, eat, nor drink any fluids. If the casualty has been bitten on an extremity, DO NOT elevate the limb; keep the extremity level with the body. Keep the casualty comfortable and reassure him. If the casualty is alone when bitten, he should go to the medical facility himself rather than wait for someone to find him. Unless the snake has been positively identified, attempt to kill it and send it with the casualty. Be sure that

Figure 6-7: Characteristics of poisonous snake bite

retrieving the snake does not endanger anyone or delay transporting the casualty.

* (1) If the bite is on an arm or leg, place a constricting band(narrow cravat [swathe], or narrow gauze bandage) one to two
finger widths above and below the bite (Figure 6-8). However, if only one constricting band is available, place that
band on the extremity between the bite site and the casualty's heart. If the bite is on the hand or foot, place a single
band above the wrist or ankle. The band should be tight
enough to stop the flow of blood near the skin, but not
tight enough to interfere with circulation. In other words,
it should not have a tourniquet-like affect. If no swelling is
seen, place the bands about one inch from either side of the
bite. If swelling is present, put the bands on the unswollen
part at the edge of the swelling. If the swelling extends
beyond the band, move the band to the new edge of the
swelling. (If possible, leave the old band on, place a new
one at the new edge of the swelling, and then remove and
save the old one in case the process has to be repeated.)
If possible, place an ice bag over the area of the bite. DO
NOT wrap the limb in ice or put ice directly on the skin.
Cool the bite area—do not freeze it. DO NOT stop to look
for ice if it will delay evacuation and medical treatment.

⚠ CAUTION
DO NOT attempt to cut open the bite nor suck out the
venom. If the venom should seep through any damaged or lacerated tissues in your mouth, you could
immediately lose consciousness or even die.

(2) If the bite is located on an arm or leg, immobilize it at a
level below the heart. DO NOT elevate an arm or leg even
with or above the level of the heart.

⚠ CAUTION
When a splint is used to immobilize the arm or leg,
take EXTREME care to ensure the splinting is done
properly and does not bind. Watch it closely and adjust it if any changes in swelling occur.

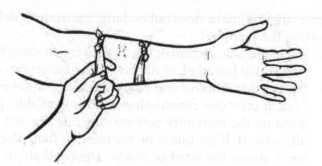

Figure 6-8: Constricting band

(3) When possible, clean the area of the bite with soap and water. DO NOT use ointments of any kind.

(4) NEVER give the casualty food, alcohol, stimulants(coffee or tea), drugs, or tobacco.

(5) Remove rings, watches, or other jewelry from the affected limb.

✍ NOTE

It may be possible, in some cases, for an aidman who is specially trained and is authorized to carry and use antivenin to administer it. The use of antivenin presents special risks, and only those with specialized training should attempt to use it!

---

d. Prevention. Except for a few species, snakes tend to be shy or passive. Unless they are injured, trapped, or disturbed, snakes usually avoid contact with humans. The harmless species are often more prone to attack. All species of snakes are usually aggressive during their breeding season.

(1) Land snakes. Many snakes are active during the period from twilight to daylight. Avoid walking as much as possible during this time.

   • Keep your hands off rock ledges where snakes are likely to be sunning.

   • Look around carefully before sitting down, particularly if in deep grass among rocks.

   • Attempt to camp on clean, level ground. Avoid camping near piles of brush, rocks, or other debris.

- Sleep on camping cots or anything that will keep you off the ground. Avoid sleeping on the ground if at all possible.
- Check the other side of a large rock before stepping over it. When looking under any rock, pull it toward you as you turn it over so that it will shield you in case a snake is beneath it.
- Try to walk only in open areas. Avoid walking close to rock walls or similar areas where snakes may be hiding.
- Determine when possible what species of snakes are likely to be found in an area which you are about to enter.
- Hike with another person. Avoid hiking alone in a snake-infested area. If bitten, it is important to have at least one companion to perform lifesaving first aid measures and to kill the snake. Providing the snake to medical personnel will facilitate both identification and treatment.
- Handle freshly killed venomous snakes only with along tool or stick. Snakes can inflict fatal bites by reflex action even after death.
- Wear heavy boots and clothing for some protection from snakebite. Keep this in mind when exposed to hazardous conditions.
- Eliminate conditions under which snakes thrive: brush, piles of trash, rocks, or logs and dense undergrowth. Controlling their food (rodents, small animals) as much as possible is also good prevention.

(2) Sea snakes. Sea snakes may be seen in large numbers but are not known to bite unless handled. Be aware of the areas where they are most likely to appear and be especially alert when swimming in these areas. Avoid swimming alone whenever possible.

WARNING

All species of snakes can swim. Many can remain under water for long periods. A bite sustained in water is just as dangerous as one on land.

6-3. Human and Other Animal Bites

Human or other land animal bites may cause lacerations or bruises. In addition to damaging tissue, human or bites from animals such as dogs, cats, bats, raccoons, or rats always present the possibility of infection.

a. **Human Bites.** Human bites that break the skin may become seriously infected since the mouth is heavily contaminated with bacteria. All human bites MUST be treated by medical personnel.

b. **Animal Bites.** Land animal bites can result in both infection and disease. Tetanus, rabies, and various types of fevers can follow an untreated animal bite. Because of these possible complications, the animal causing the bite should, if possible, be captured or killed (without damaging its head) so that competent authorities can identify and test the animal to determine if it is carrying diseases.

c. **First Aid.**

  (1) Cleanse the wound thoroughly with soap or detergent solution.

  (2) Flush it well with water.

  (3) Cover it with a sterile dressing.

  (4) Immobilize an injured arm or leg.

  (5) Transport the casualty immediately to a medical treatment facility.

✍ NOTE

If unable to capture or kill the animal, provide medical personnel with any information possible that will help identify it. Information of this type will aid in appropriate treatment.

6-4. Marine (Sea) Animals

With the exception of sharks and barracuda, most marine animals will not deliberately attack. The most frequent injuries from marine animals are wounds by biting, stinging, or puncturing. Wounds inflicted by marine animals can be very painful, but are rarely fatal.

a. **Sharks, Barracuda, and Alligators.** Wounds from these marine animals can involve major trauma as a result of bites and lacerations. Bites from large marine animals are potentially the most life threatening of all injuries from marine animals. Major wounds from these animals can be treated by controlling the bleeding, preventing shock, giving basic life support, splinting the injury, and by securing prompt medical aid.

b. Turtles, Moray Eels, and Corals. These animals normally inflict minor wounds. Treat by cleansing the wound(s) thoroughly and by splinting if necessary.

c. Jellyfish, Portuguese men-of-war, Anemones, and Others. This group of marine animals inflict injury by means of stinging cells in their tentacles. Contact with the tentacles produces burning pain with a rash and small hemorrhages on the skin. Shock, muscular cramping, nausea, vomiting, and respiratory distress may also occur. Gently remove the clinging tentacles with a towel and wash or treat the area. Use diluted ammonia or alcohol, meat tenderizer, and talcum powder. If symptoms become severe or persist, seek medical aid.

d. Spiny Fish, Urchins, Stingrays, and Cone Shells. These animals inject their venom by puncturing with their spines. General signs and symptoms include swelling, nausea, vomiting, generalized cramps, diarrhea, muscular paralysis, and shock. Deaths are rare. Treatment consists of soaking the wounds in hot water (when available) for 30 to 60 minutes. This inactivates the heat sensitive toxin. In addition, further first aid measures (controlling bleeding, applying a dressing, and so forth) should be carried out as necessary.

> ### ⚠ CAUTION
> Be careful not to scald the casualty with water that is too hot because the pain of the wound will mask the normal reaction to heat.

6-5. Insect Bites/Stings

An insect bite or sting can cause great pain, allergic reaction, inflammation, and infection. If not treated correctly, some bites/stings may cause serious illness or even death. When an allergic reaction is not involved, first aid is a simple process. In any case, medical personnel should examine the casualty at the earliest possible time. It is important to properly identify the spider, bee, or creature that caused the bite/sting, especially in cases of allergic reaction when death is a possibility.

a. Types of Insects. The insects found throughout the world that can produce a bite or sting are too numerous to mention in detail. Commonly encountered stinging or biting insects include brown recluse spiders (Figure 6-9), black widow spiders (Figure 6-10), tarantulas (Figure 6-11), scorpions (Figure 6-12), urticating caterpillars, bees, wasps, centipedes, conenose beetles (kissing bugs), ants, and wheel bugs. Upon being reassigned, especially to overseas areas, take the time to become acquainted with the types of insects to avoid.

b. Signs/Symptoms. Discussed in paragraphs (1) and (2) below are the most common effects of insect bites/stings. They can occur alone or in combination with the others.

(1) Less serious. Commonly seen signs/symptoms are pain, irritation, swelling, heat, redness, and itching. Hives or wheals (raised areas of the skin that itch) may occur. These are the least severe of the allergic reactions that commonly occur from insect bites/stings. They are usually dangerous only if they affect the air passages (mouth, throat, nose, and so forth), which could interfere with breathing. The bites/stings of bees, wasps, ants, mosquitoes, fleas, and ticks are usually not serious and normally produce mild and localized symptoms. A tarantula's bite is usually

Figure 6-9: Brown recluse spider

Figure 6-10: Black widow spider

Figure 6-11: Tarantula

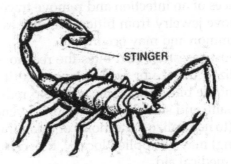

STINGER

Figure 6-12: Scorpion

no worse than that of a bee sting. Scorpions are rare and their stings (except for a specific species found only in the Southwest desert) are painful but usually not dangerous.

(2) Serious. Emergency allergic or hypersensitive reactions sometimes result from the stings of bees, wasps, and ants. Many people are allergic to the venom of these particular insects. Bites or stings from these insects may produce more serious reactions, to include generalized itching and hives, weakness, anxiety, headache, breathing difficulties, nausea, vomiting, and diarrhea. Very serious allergic reactions (called anaphylactic shock) can lead to complete collapse, shock, and even death. Spider bites (particularly from the black widow and brown recluse spiders) can be serious also. Venom from the black widow spider affects the nervous system. This venom can cause muscle cramps, a rigid, nontender abdomen, breathing difficulties, sweating, nausea and vomiting. The brown recluse spider generally produces local rather than system-wide

problems; however, local tissue damage around the bite can be severe and can lead to an ulcer and even gangrene.

c. First Aid. There are certain principles that apply regardless of what caused the bite/sting. Some of these are:

- If there is a stinger present, for example, from a bee, remove the stinger by scraping the skin's surface with a fingernail or knife. DO NOT squeeze the sac attached to the stinger because it may inject more venom.
- Wash the area of the bite/sting with soap and water (alcohol or an antiseptic may also be used) to help reduce the chances of an infection and remove traces of venom.
- Remove jewelry from bitten extremities because swelling is common and may occur.
- In most cases of insect bites the reaction will be mild and localized use ice or cold compresses (if available) on the site of the bite/sting. This will help reduce swelling, ease the pain, and slow the absorption of venom. Meat tenderizer (to neutralize the venom) or calamine lotion (to reduce itching) may be applied locally. If necessary,
- seek medical aid.
- In more serious reactions (severe and rapid swelling, allergic symptoms, and so forth) treat the bite/sting like you would treat a snakebite; that is, apply constricting bands above and below the site. See paragraph 6-2c(1) above for details and illustration (Figure 6-8) of a constricting band.
- * Be prepared to perform basic lifesaving measures, such as rescue breathing. Reassure the casualty and keep him calm.
- In serious reactions, attempt to capture the insect for positive identification; however, be careful not to become a casualty yourself.
- If the reaction or symptoms appear serious, seek medical aid immediately.

⚠ *CAUTION

Insect bites/stings may cause anaphylactic shock (a shock caused by a severe allergic reaction). This is a life-threatening event and a MEDICAL EMERGENCY! Be prepared to immediately transport the casualty to a medical facility.

✍ NOTE

Be aware that some allergic or hypersensitive individuals may carry identification (such as a MEDIC ALERT tag) or emergency insect bite treatment kits. If the casualty is having an allergic reaction and has such a kit, administer the medication in the kit according to the instructions which accompany the kit.

---

d. Prevention. Some prevention principles are:
- Apply insect repellent to all exposed skin, such as the ankles to prevent insects from creeping between uniform and boots. Also apply the insect repellent to the shoulder blades where the shirt fits tight enough that mosquitoes bite through. DO NOT apply insect repellent to the eyes.
- Reapply repellent, every 2 hours during strenuous activity and soon after stream crossings.
- Blouse the uniform inside the boots to further reduce risk.
- Wash yourself daily if the tactical situation permits. Pay particular attention to the groin and armpits.
- Use the buddy system. Check each other for insect bites.
- Wash your uniform at least weekly.

6-6. Table See Table 6-1 for information on bites and stings.

Table 6-1: Bites and Stings

TYPES	FIRST AID
Snakebite	1. Move the casualty away from the snake.
	2. Remove all rings and bracelets from the affected extremity.
	3. Reassure the casualty and keep him quiet.
	4. Place ice or freeze pack, if available, over the area of the bite.
	5. Apply constricting band(s) 1-2 finger widths from the bite. One should be able to insert a finger between the band and the skin.
	• Arm or leg bite—place one band above and one band below the bite site.
	• Hand or foot bite—place one band above the wrist or ankle.

(continued)

Table 6-1: *(Continued)*

TYPES	FIRST AID
	6. Immobilize the affected part in a position below the level of the heart.
	7. Kill the snake (if possible, without damaging its head or endangering yourself) and send it with the casualty.
	8. Seek medical aid immediately.
Brown Recluse Spider or Black Widow Spider Bite	1. Keep the casualty quiet. 2. Wash the area. 3. Apply ice or freeze pack, if available. 4. Seek medical aid.
Tarantula Bite or Scorpion Sting or Ant Bites	1. Wash the area. 2. Apply ice or freeze pack, if available. 3. Apply baking soda, calamine lotion, or meat tenderizer to bite site to relieve pain and itching. 4. If site of bite(s) or sting(s) is on the face, neck (possible airway problems), or genital area, or if local reaction seems severe, or if the sting is by the dangerous type of scorpion found in the Southwest desert, keep the casualty quiet as possible and seek immediate medical aid.
Bee Stings	1. If the stinger is present, remove by scraping with a knife or fingernail. DO NOT squeeze venom sac on stinger; more venom may be injected. 2. Wash the area. 3. Apply ice or freeze pack, if available. ★4. If allergic signs/symptoms appear, be prepared to seek immediate medical aid.

CHAPTER 7

FIRST AID IN TOXIC ENVIRONMENTS

American forces have not been exposed to high levels of toxic substances on the battlefield since World War I. In future conflicts and wars we can expect the use of such agents. Chemical weapons will degrade unit effectiveness rapidly by forcing troops to wear hot protective clothing and by creating confusion and fear. Through training in protective procedures and first aid, units can maintain their effectiveness on the integrated battlefield.

SECTION I. INDIVIDUAL PROTECTION AND FIRST AID EQUIPMENT FOR TOXIC SUBSTANCES

7-1. Toxic Substances

a. Gasoline, chlorine, and pesticides are examples of common toxic substances. They may exist as solids, liquids, or gases depending upon temperature and pressure. Gasoline, for example, is a vaporizable liquid; chlorine is a gas; and Warfarin, a pesticide, is a solid. Some substances are more injurious to the body than others when they are inhaled or eaten or when they contact the skin or eyes. Whether they are solids, liquids, or gases (vapors and aerosols included), they may irritate, inflame, blister, burn, freeze, or destroy tissue such as that associated with the respiratory tract or the eyes. They may also be absorbed into the bloodstream, disturbing one or several of the body's major functions.

b. You may come in contact with toxic substances in combat or in everyday activities. Ordinarily, brief exposures to common household toxic substances, such as disinfectants and bleach solutions, do not cause injuries. Exposure to toxic chemical agents in warfare, even for a few seconds, could result in death, injury, or incapacitation. Remember that toxic substances

179

employed by an enemy could persist for hours or days. To survive and operate effectively in a toxic environment, you must be prepared to protect yourself from the effects of chemical agents and to provide first aid to yourself and to others.

7-2. Protective and First Aid Equipment

You are issued equipment for protection and first aid treatment in a toxic environment. You must know how to use the items described in *a* through *e*. It is equally important that you know when to use them. Use your protective clothing and equipment when you are ordered to and when you are under a nuclear, biological, or chemical (NBC) attack. Also, use your protective clothing and equipment when you enter an area where NBC agents have been employed.

a. Field Protective Mask With Protective Hood. Your field protective mask is the most important piece of protective equipment. You are given special training in its use and care.

b. Field Protective Clothing. Each soldier is authorized three sets of the following field protective clothing:
   • Overgarment ensemble (shirt and trousers), chemical protective.
   • Footwear cover (overboots), chemical protective.
   • Glove set, chemical protective.

c. Nerve Agent Pyridostigmine Pretreatment (NAPP). You will be issued a blister pack of pretreatment tablets when your commander directs. When ordered to take the pretreatment you must take one tablet every eight hours. This must be taken prior to exposure to nerve agents, since it may take several hours to develop adequate blood levels.

> #### ✍ NOTE
> Normally, one set of protective clothing is used in acclimatization training that uses various mission-oriented protective posture (MOPP) levels.

d. M258A1 Skin Decontamination Kit. The M258A1 Skin Decontamination (decon) Kit contains three each of the following:
   • DECON-1 packets containing wipes (pads) moistened with decon solution.

- DECON-2 packets containing dry wipes (pads) previously moistened with decon solution and sealed glass ampules. Ampules are crushed to moisten pads.

> **WARNING**
>
> The decon solution contained in both DE-CON-1 and DECON-2 packets is a poison and caustic hazard and can permanently damage the eyes. Keep wipes out of the eyes, mouth, and open wounds. Use WATER to wash toxic agent out of eyes and wounds and seek medical aid.

e. Nerve Agent Antidote Kit, Mark I (NAAK MKI). Each soldier is authorized to carry three Nerve Agent Antidote Kits, Mark I, to treat nerve agent poisoning. When NAPP has been taken several hours (but no greater than 8 hours) prior to exposure, the NAAK MKI treatment of nerve agent poisoning is much more effective.

SECTION II. CHEMICAL-BIOLOGICAL AGENTS

7-3. Classification

a. Chemical agents may be classified according to the primary physiological effects they produce, such as nerve, blister, blood, choking, vomiting, and incapacitating agents.
b. Biological agents may be classified according to the effect they have on man. These include blockers, inhibitors, hybrids, and membrane active compounds. These agents are found in living organisms such as fungi, bacteria and viruses.

> **WARNING**
>
> Ingesting water or food contaminated with nerve, blister, and other chemical agents and with some biological agents can be fatal. NEVER consume water or food which is suspected of being contaminated until it has been tested and found safe for consumption.

7-4. Conditions for Masking Without Order or Alarm

Once an attack with a chemical or biological agent is detected or suspected, or information is available that such an agent is about to be used, you must STOP breathing and mask immediately. DO NOT WAIT to receive an order or alarm under the following circumstances:

- Your position is hit by artillery or mortar fire, missiles, rockets, smokes, mists, aerial sprays, bombs, or bomblets.
- Smoke from an unknown source is present or approaching.
- A suspicious odor, liquid, or solid is present.
- A toxic chemical or biological attack is present.
- You are entering an area known or suspected of being contaminated.
- During any motor march, once chemical warfare has begun.
- When casualties are being received from an area where chemical or biological agents have reportedly been used.
- You have one or more of the following symptoms:
  - An unexplained runny nose.
  - A feeling of choking or tightness in the chest or throat.
  - Dimness of vision.
  - Irritation of the eyes.
  - Difficulty in or increased rate of breathing without
  - Sudden feeling of depression.
  - Dread, anxiety, restlessness.
  - Dizziness or light-headedness.
  - Slurred speech.
- Unexplained laughter or unusual behavior is noted in others.
- Numerous unexplained ill personnel.
- Buddies suddenly collapsing without evident cause.
- Animals or birds exhibiting unusual behavior and/or sudden unexplained death.

7-5. First Aid for a Chemical Attack (081-831-1030 and 081-831-1031)

Your field protective mask gives protection against chemical as well as biological agents. Previous practice enables you to mask in 9 seconds or less or to put on your mask with hood within 15 seconds.

a. Step ONE (081-831-1030 and 081-831-1031). Stop breathing. Don your mask, seat it properly, clear and check your mask, and resume breathing. Give the alarm, and continue the mission. Keep your mask on until the "all clear" signal has been given.

✍ NOTE

Keep your mask on until the area is no longer hazardous and you are told to unmask.

---

b. Step TWO (081-831-1030).If symptoms of nerve agent poisoning (paragraph 7-7) appear, immediately give yourself a nerve agent antidote. You should have taken NAPP several hours prior to exposure which will enhance the action of the nerve agent antidote.

△ CAUTION

Do not inject a nerve agent antidote until you are sure you need it.

---

c. Step THREE (081-831-1031). If your eyes and face become contaminated, you must immediately try to get under cover. You need this shelter to prevent further contamination while performing decon procedures on areas of the head. If no overhead cover is available, throw your poncho or shelter half over your head before beginning the decon process. Then you should put on the remaining protective clothing. (See Appendix F for decon procedure.) If vomiting occurs, the mask should be lifted momentarily and drained—while the eyes are closed and the breath is held—and replaced, cleared, and sealed.

d. Step FOUR. If nerve agents are used, mission permitting, watch for persons needing nerve agent antidotes and immediately follow procedures outlined in paragraph 7-8 b.

e. STEP FIVE. When your mission permits, decon your clothing and equipment.

SECTION III. NERVE AGENTS

7-6. Background Information

a. Nerve agents are among the deadliest of chemical agents. They can be delivered by artillery shell, mortar shell, rocket, missile, landmine, and aircraft bomb, spray, or bomblet. Nerve agents enter the body by inhalation, by ingestion, and through the skin. Depending on the route of entry and the amount, nerve agents can produce injury or death within minutes. Nerve agents also can achieve their effects with small amounts. Nerve agents are absorbed rapidly, and the effects are felt immediately upon entry into the body. You will be issued three Nerve Agent Antidote Kits, Mark I. Each kit consists of one atropine autoinjector and one pralidoxime chloride (2 PAM Cl) auto injector (also called injectors) (Figure 7-1).

b. When you have the signs and symptoms of nerve agent poisoning, you should immediately put on the protective mask and then inject yourself with one set of the Nerve Agent Antidote Kit, Mark I. You should inject yourself in the outside (lateral) thigh muscle or if you are thin, in the upper outer (lateral) part of the buttocks.

c. Also, you may come upon an unconscious chemical agent casualty who will be unable to care for himself and who will require your aid. You should be able to successfully—

(1) Mask him if he is unmasked.

(2) Inject him, if necessary, with all his autoinjectors.

(3) Decontaminate his skin.

(4) Seek medical aid.

7-7. Signs/Symptoms of Nerve Agent Poisoning (081-831-1030 and 081-831-1031)

The symptoms of nerve agent poisoning are grouped as MILD—those which you recognize and for which you can perform self-aid, and SEVERE—those which require buddy aid.

a. MILD Symptoms (081-831-1030).

• Unexplained runny nose.

• Unexplained sudden headache.

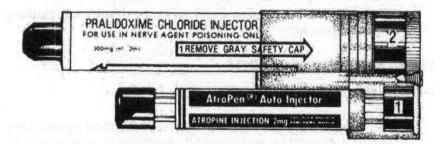

Figure 7-1: Nerve Agent Antidote Kit, Mark I.

- Sudden drooling.
- Difficulty seeing (blurred vision).
- Tightness in the chest or difficulty in breathing.
- Localized sweating and twitching (as a result of small amount of nerve agent on skin).
- Stomach cramps.
- Nausea.

b. SEVERE Signs/Symptoms (081-831-1031).
  - Strange or confused behavior.
  - Wheezing, difficulty in breathing, and coughing.
  - Severely pinpointed pupils.
  - Red eyes with tearing (if agent gets into the eyes).
  - Vomiting.
  - Severe muscular twitching and general weakness.
  - Loss of bladder/bowel control.
  - Convulsions.
  - Unconsciousness.
  - Stoppage of breathing.

7-8. First Aid for Nerve Agent Poisoning (081-831-1030) and (081-831-1031)

The injection site for administering the Nerve Agent Antidote Kit, Mark I (see Figure 7-1), is normally in the outer thigh muscle (see Figure 7-2). It is important that the injections be given into a large muscle area. If the individual is thinly-built, then the injections must be administered into the upper outer quarter (quadrant) of the buttocks (see Figure 7-3). This avoids injury to the thigh bone.

> **WARNING**
>
> There is a nerve that crosses the buttocks, so it is important to inject only into the upper outer quadrant (see Figure 7-3). This will avoid injuring this nerve. Hitting the nerve can cause paralysis.

a. Self-Aid (081-831-1030).
  (1) Immediately put on your protective mask after identifying any of the signs/symptoms of nerve agent poisoning (paragraph 7-7).
  (2) Remove one set of the Nerve Agent Antidote Kit, Mark I.

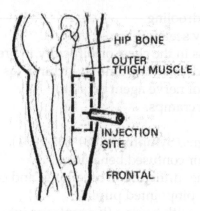

Figure 7-2: Thigh injection site

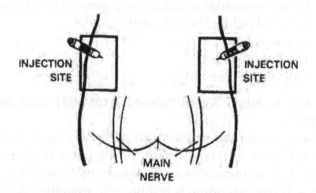

Figure 7-3: Buttocks injections site

(3) With your non dominant hand, hold the autoinjectors by the plastic clip so that the larger autoinjector is on top and both are positioned in front of you at eye level (see Figure 7-4).

(4) With the other hand, check the injection site (thigh or buttocks) for buttons or objects in pockets which may interfere with the injections.

(5) Grasp the atropine (smaller) autoinjector with the thumb and first two fingers (see Figure 7-5).

⚠ CAUTION

DO NOT cover/hold the green (needle) end with your hand or fingers—you might accidentally inject yourself.

(6) Pull the injector out of the clip with a smooth motion(see Figure 7-6).

WARNING
The injector is now armed. DO NOT touch the green (needle) end.

(7) Form a fist around the autoinjector. BE CAREFUL NOT TO INJECT YOURSELF IN THE HAND!

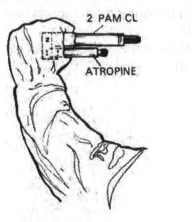

Figure 7-4: Holding the set of autoinjectors by the plastic clip

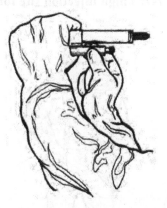

Figure 7-5: Grasping the atropine autoinjector between the thumb and first two fingers of the hand

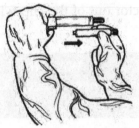

Figure 7-6: Removing the atropine autoinjector from the clip

(8) Position the green end of the atropine autoinjector against the injection site (thigh or buttocks):

    (a) On the outer thigh muscle (see Figure 7-7).
       OR

    (b) On the upper outer portion of the buttocks (see Figure 7-8).

Figure 7-7: Thigh injection site for self-aid

Figure 7-8: Buttocks injection site for self-aid

(9) Apply firm, even pressure (not a jabbing motion) to the injector until it pushes the needle into your thigh (or buttocks).

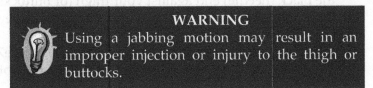

WARNING

Using a jabbing motion may result in an improper injection or injury to the thigh or buttocks.

✍ **NOTE**

Firm pressure automatically triggers the coiled spring mechanism. This plunges the needle through the clothing into the muscle and injects the fluid into the muscle tissue.

---

(10) Hold the injector firmly in place for at least ten seconds. The ten seconds can be estimated by counting "one thousand and one, one thousand and two," and so forth.
(11) Carefully remove the autoinjector.
(12) Place the used atropine injector between the little finger and the ring finger of the hand holding the remaining autoinjector and the clip (see Figure 7-9). WATCH OUT FOR THE NEEDLE!

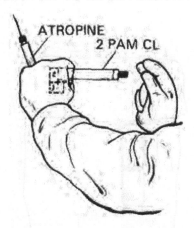

Figure 7-9: Used atropine autoinjector placed between the little finger and ring finger

(13) Pull the 2 PAM C1 autoinjector (the larger of the two injectors) out of the clip (see Figure 7-10) and inject yourself in the same manner as steps (7) through (11) above, holding the black (needle) end against your thigh (or buttocks).

(14) Drop the empty injector clip without dropping the used autoinjectors.

(15) Attach the used injectors to your clothing (see Figure7-11). Be careful NOT to tear your protective gloves/clothing with the needles.

    (a) Push the needle of each injector (one at a time) through one of the pocket flaps of your protective overgarment.

    (b) Bend each needle to form a hook.

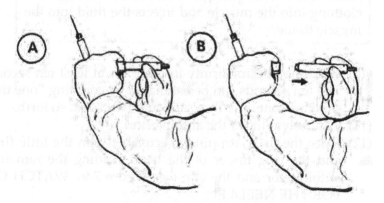

Figure 7-10: Removing the 2 PAM Cl autoinjector

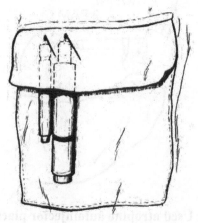

Figure 7-11: One set of used autoinjectors attached to pocket flap

> **WARNING**
>
> It is important to keep track of all used auto-injectors so that medical personnel can determine how much antidote has been given and the proper follow-up treatment can be provided, if needed.

(16) Massage the injection site if time permits.

> **WARNING**
>
> If within 5 to 10 minutes after administering the first set of injections, your heart begins to beat rapidly and your mouth becomes very dry, DO NOT give yourself another set of injections. You have already received enough antidote to overcome the dangerous effects of the nerve agent. If you are able to walk without assistance (ambulate), know who you are and where you are, you WILL NOT need the second set of injections. (If not needed, giving yourself a second set of injections may create a nerve agent antidote overdose, which could cause incapacitation.) If, however, you continue to have symptoms of nerve agent poisoning for 10 to 15 minutes after receiving one set of injections, seek a buddy to check your symptoms. If your buddy agrees that your symptoms are worsening, administer the second set of injections.

> **✍ NOTE (081-831-1030)**
>
> While waiting between sets (injections), you should decon your skin, if necessary, and put on the remaining protective clothing.

b. Buddy aid (081-831-1031).
   A soldier exhibiting SEVERE signs/symptoms of nerve agent poisoning will not be able to care for himself and must therefore be given buddy aid as quickly as possible. Buddy aid will

be required when a soldier is totally and immediately inca-pacitated prior to being able to apply self-aid, and all three sets of his Nerve Agent Antidote Kit, Mark I, need to be given by a buddy. Buddy aid may also be required after a soldier attempted to counter the nerve agent by self-aid but became incapacitated after giving himself one set of the autoinjectors. Before initiating buddy aid, a buddy should determine if one set of injectors has already been used so that no more than three sets of the antidote are administered.

(1) Move (roll) the casualty onto his back (face up) if not already in that position.

WARNING
Avoid unnecessary movement of the casualty so as to keep from spreading the contamination.

(2) Remove the casualty's protective mask from the carrier.
(3) Position yourself above the casualty's head, facing his feet.

WARNING
Squat, DO NOT kneel, when masking a chemical agent casualty. Kneeling may force the chemical agent into or through your protective clothing, which will greatly reduce the effectiveness of the clothing.

(4) Place the protective mask on the casualty.
(5) Have the casualty clear the mask.
(6) Check for a complete mask seal by covering the inlet valves. If properly sealed the mask will collapse.

NOTE
If the casualty is unable to follow instructions, is unconscious, or is not breathing, he will not be able to perform steps (5) or (6). It may, therefore, be impossible to determine if the mask is sealed. But you should still try to check for a good seal by placing your hands over the valves.

(7) Pull the protective hood over the head, neck, and shoulders of the casualty.

(8) Position yourself near the casualty's thigh.

(9) Remove one set of the casualty's autoinjectors.

> ✍ **NOTE (081-831-1031)**
> Use the CASUALTY'S autoinjectors. DO NOT use YOUR autoinjectors for buddy aid; if you do, you may not have any antidote if/when needed for self-aid.

(10) With your nondominant hand, hold the set of autoinjectors by the plastic clip so that the larger autoinjector is on top and both are positioned in front of you at eye level (see Figure 7-4).

(11) With the other hand, check the injection site (thigh or buttocks) for buttons or objects in pockets which may interfere with the injections.

(12) Grasp the atropine (smaller) autoinjector with the thumb and first two fingers (see Figure 7-5).

> ⚠ **CAUTION**
> DO NOT cover/hold the green (needle) end with your hand or fingers–you may accidentally inject yourself.

(13) Pull the injector out of the clip with a smooth motion (see Figure 7-6).

> **WARNING**
> The injector is now armed. DO NOT touch the green (needle) end.

(14) Form a fist around the autoinjector. BE CAREFUL NOT TO INJECT YOURSELF IN THE HAND.

> **WARNING**
> Holding or covering the needle (green) end of the autoinjector may result in accidentally injecting yourself.

(15) Position the green end of the atropine autoinjector against the injection site (thigh or buttocks):
   (a) On the casualty's outer thigh muscle (see Figure 7-12).

✍ **NOTE**

The injections are normally given in the casualty's thigh.

WARNING

If this is the injection site used, be careful not to inject him close to the hip, knee, or thigh bone.

OR

   (b) On the upper outer portion of the casualty's buttocks (see Figure 7-13).

✍ **NOTE**

If the casualty is thinly built, reposition him onto his side or stomach and inject the antidote into his buttocks.

Figure 7-12: Injecting the casualty's thigh

WARNING

Inject the antidote only into the upper outer portion of his buttocks (see Figure 7-13). This avoids hitting the nerve that crosses the buttocks. Hitting this nerve can cause paralysis.

(16) Apply firm, even pressure (not a jabbing motion) to the injector to activate the needle. This causes the needle to penetrate both the casualty's clothing and muscle.

WARNING

Using a jabbing motion may result in an improper injection or injury to the thigh or buttocks.

(17) Hold the injector firmly in place for at least ten seconds. The ten seconds can be estimated by counting "one thousand and one, one thousand and two, " and so forth.

(18) Carefully remove the autoinjector.

Figure 7-13: Injecting the casualty's buttocks

(19) Place the used autoinjector between the little finger and ring finger of the hand holding the remaining autoinjector and the clip(see Figure 7-9). WATCH OUT FOR THE NEEDLE!

(20) Pull the 2 PAM Cl autoinjector (the larger of the two injectors) out of the clip (see Figure 7-10) and inject the casualty in the same manner as steps (9) through (19) above, holding the black (needle) end against the casualty's thigh (or buttocks).

(21) Drop the clip without dropping the used autoinjectors.

(22) Carefully lay the used injectors on the casualty's chest (if he is lying on his back), or on his back (if he is lying on his stomach), pointing the needles toward his head.

(23) Repeat the above procedure immediately (steps 9through 22), using the second and third set of autoinjectors.

(24) Attach the three sets of used autoinjectors to the casualty's clothing (see Figure 7-14). Be careful NOT to tear either your or the casualty's protective clothing/gloves with the needles.

  (a) Push the needle of each injector (one at a time) through one of the pocket flaps of his protective overgarment.

  (b) Bend each needle to form a hook.

> **WARNING**
> It is important to keep track of all used autoinjectors so that medical personnel will be able to determine how much antidote has been given and the proper follow-up/treatment can be provided, if needed.

(25) Massage the area if time permits.

SECTION IV. OTHER AGENTS

7-9. Blister Agents

Blister agents (vesicants) include mustard (HD), nitrogen mustards (HN), lewisite (L), and other arsenicals, mixtures of mustards and arsenical, and phosgene oxime (CX). Blister agents act on the eyes, mucous membranes, lungs, and skin. They burn and blister the skin or any other body parts they contact. Even relatively low doses may cause serious injury. Blister agents damage the respiratory tract (nose,

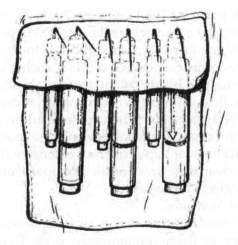

Figure 7-14: Three sets of used autoinjectors attached to pocket flap

sinuses and windpipe) when inhaled and cause vomiting and diarrhea when absorbed. Lewisite and phosgene oxime cause immediate
pain on contact. However, mustard agents are deceptive and there is
little or no pain at the time of exposure. Thus, in some cases, signs of
injury may not appear for several hours after exposure.

a. Protective Measures. Your protective mask with hood and
   protective overgarments provide you protection against blister agents. If it is known or suspected that blister agents are
   being used, STOP BREATHING, put on your mask and all
   your protective overgarments.

> **⚠ CAUTION**
> Large drops of liquid vesicants on the protective over-
> garment ensemble may penetrate it if allowed to stand
> for an extended period. Remove large drops as soon
> as possible.

b. Signs/Symptoms of Blister Agent Poisoning.
   (1) Immediate and intense pain upon contact (lewisite and phos-
       gene oxime). No initial pain upon contact with mustard.
   (2) Inflammation and blisters (burns)–tissue destruction.
       The severity of a chemical burn is directly related to the
       concentration of the agent and the duration of contact

with the skin. The longer the agent is in contact with the tissue, the more serious the injury will be.

(3) Vomiting and diarrhea. Exposure to high concentrations of vesicants may cause vomiting anchor diarrhea.

(4) Death. The blister agent vapors absorbed during ordinary field exposure will probably not cause enough internal body (systemic) damage to result in death. However, death may occur from prolonged exposure to high concentrations of vapor or from extensive liquid contamination over wide areas of the skin, particularly when decon is neglected or delayed.

c. First Aid Measures.

(1) Use your M258A1 decon kit to decon your skin and use water to flush contaminated eyes. Decontamination of vesicants must be done immediately (within 1 minute is best).

(2) If blisters form, cover them loosely with a field dressing and secure the dressing.

⚠ CAUTION

Blisters are actually burns. DO NOT attempt to decon the skin where blisters have formed.

(3) If you receive blisters over a wide area of the body, you are considered seriously burned. SEEK MEDICAL AID IMMEDIATELY.

(4) If vomiting occurs, the mask should be lifted momentarily and drained—while the eyes are closed and the breath is held–and replaced, cleared, and sealed.

(5) Remember, if vomiting or diarrhea occurs after having been exposed to blister agents, SEEK MEDICAL AID IMMEDIATELY.

7-10. Choking Agents (Lung-Damaging Agents)

Chemical agents that attack lung tissue, primarily causing fluid build-up (pulmonary edema), are classified as choking agents (lung-damaging agents). This group includes phosgene (CG), diaphosgene (DP), chlorine (CL), and chloropicrin (PS). Of these four agents, phosgene is the most dangerous and is more likely to be employed by the enemy in future conflict.

a. Protective Measures. Your protective mask gives adequate protection against choking agents.

b. Signs/Symptoms. During and immediately after exposure to choking agents (depending on agent concentration and length of exposure), you may experience some or all of the following signs/symptoms:

- Tears (lacrimation).
- Dry throat.
- Coughing.
- Choking.
- Tightness of chest.
- Nausea and vomiting.
- Headaches.

c. First Aid Measures.

(1) If you come in contact with phosgene, your eyes become irritated, or a cigarette becomes tasteless or offensive, STOP BREATHING and put on your mask immediately.

(2) If vomiting occurs, the mask should be lifted momentarily and drained—while the eyes are closed and the breath is held–replaced, cleared, and sealed.

(3) Seek medical assistance if any of the above signs/symptoms occur.

✍ NOTE

If you have no difficulty breathing, do not feel nauseated, and have no more than the usual shortness of breath on exertion, then you inhaled only a minimum amount of the agent. You may continue normal duties.

---

d. Death. With ordinary field exposure to choking agents, death will probably not occur. However, prolonged exposure to high concentrations of the vapor and neglect or delay in masking can be fatal.

7-11. Blood Agents

Blood agents interfere with proper oxygen utilization in the body. Hydrogen cyanide (AC) and cyanogen chloride (CK) are the primary agents in this group.

a. Protective Measures. Your protective mask with a fresh filter gives adequate protection against field concentrations of blood agent vapor. The protective overgarment as well as the mask are needed when exposed to liquid hydrogen cyanide.

b. Signs/Symptoms. During and immediately after exposure to blood agents (depending on agent concentration and length of exposure), you may experience some or all of the following signs/symptoms:

   • Eye irritation.
   • Nose and throat irritation.
   • Sudden stimulation of breathing.
   • Nausea.
   • Coughing.
   • Tightness of chest.
   • Headache.
   • Unconsciousness.

c. First Aid Measures.

   (1) Hydrogen cyanide. During any chemical attack, if you get a sudden stimulation of breathing or notice an odor like bitter almonds, PUT ON YOUR MASK IMMEDIATELY. Speed is absolutely essential since this agent acts so rapidly that within a few seconds its effects will make it impossible for individuals to put on their mask by themselves. Stop breathing until the mask is on, if at all possible. This may be very difficult since the agent strongly stimulates respiration.

   (2) Cyanogen chloride. PUT ON YOUR MASK IMMEDIATELY if you experience any irritation of the eyes, nose, or throat.

d. Medical Assistance. If you suspect that you have been exposed to blood agents, seek medical assistance immediately.

7-12. Incapacitating Agents

Generally speaking, an incapacitating agent is any compound which can interfere with your performance. The agent affects the central nervous system and produces muscular weakness and abnormal behavior. It is likely that such agents will be disseminated by smoke-producing munitions or aerosols, thus making breathing their means of entry into the body. The protective mask is, therefore, essential.

a. There is no special first aid to relieve the symptoms of incapacitating agents. Supportive first aid and physical restraint may be indicated. If the casualty is stuporous or comatose, be sure that respiration is unobstructed; then turn him on his stomach with his head to one side (in case vomiting should occur). Complete cleansing of the skin with soap and water should be done as soon as possible; or, the M258A1 Skin Decontamination Kit can be used if washing is impossible. Remove weapons and other potentially harmful items from the possession of individuals who are suspected of having these symptoms. Harmful items include cigarettes, matches, medications, and small items which might be swallowed accidentally. Delirious persons have been known to attempt to eat items bearing only a superficial resemblance to food.

  b. Anticholinergic drugs (BZ - type) may produce alarming dryness and coating of the lips and tongue; however, there is usually no danger of immediate dehydration. Fluids should be given sparingly, if at all, because of the danger of vomiting and because of the likelihood of temporary urinary retention due to paralysis of bladder muscles. An important medical consideration is the possibility of heatstroke caused by the stoppage of sweating. If the environmental temperature is above 78° F, and the situation permits, remove excessive clothing from the casualty and dampen him to allow evaporative cooling and to prevent dehydration. If he does not readily improve, apply first aid measures for heat stroke and seek medical attention.

7-13. Incendiaries
Incendiaries can be grouped as white phosphorus, thickened fuel, metal, and oil and metal. You must learn to protect yourself against these incendiaries.

  a. White phosphorus (WP) is used primarily as a smoke producer but can be used for its incendiary effect to ignite field expedients and combustible materials. The burns from WP are usually multiple, deep, and variable in size. When particles of WP get on the skin or clothing, they continue to burn until deprived of air. They also have a tendency to stick to a surface and must be brushed off or picked out.
    (1) If burning particles of phosphorus strike and stick to your clothing, quickly take off the contaminated clothing before the phosphorus burns through to the skin.
    (2) If burning phosphorus strikes your skin, smother the flame by submerging yourself in water or by dousing the

WP with water from your canteen or any other source. Urine, a wet cloth, or mud can also be used.

✍ **NOTE**

Since WP is poisonous to the system, DO NOT use grease or oil to smother the flame. The WP will be absorbed into the body with the grease or oil.

---

    (3) Keep the WP particles covered with wet material to exclude air until you can remove them or get them removed from your skin.

    (4) Remove the WP particles from the skin by brushing them with a wet cloth and by picking them out with a knife, bayonet, stick, or other available object.

    (5) Report to a medical facility for treatment as soon as your mission permits.

  b. Thickened fuel mixtures (napalm) have a tendency to cling to clothing and body surfaces, thereby producing prolonged exposure and severe burns. The first aid for these burns is the same as for other heat burns. The heat and irritating gases given off by these combustible mixtures may cause lung damage, which must be treated by a medical officer.

  c. Metal incendiaries pose special problems. Thermite and thermate particles on the skin should be immediately cooled with water and then removed. Even though thermate particles have their own oxygen supply and continue to burn under water, it helps to cool them with water. The first aid for these burns is the same as for other heat burns. Particles of magnesium on the skin burn quickly and deeply. Like other metal incendiaries, they must be removed. Ordinarily, the complete removal of these particles should be done by trained personnel at a medical treatment facility, using local anesthesia. Immediate medical treatment is required.

  d. Oil and metal incendiaries have much the same effect on contact with the skin and clothing as those discussed (b and c above). Appropriate first aid measures for burns are described in Chapter 3.

7-14. First Aid for Biological Agents

We are concerned with victims of biological attacks and with treating symptoms after the soldier becomes ill. However, we are more concerned with preventive medicine and hygienic measures taken before the attack. By accomplishing a few simple tasks we can minimize their effects.

   a. Immunizations. In the military we are accustomed to keeping inoculations up to date. To prepare for biological defense, every effort must be taken to keep immunizations current. Based on enemy capabilities and the geographic location of our operations, additional immunizations may be required.
   b. Food and Drink. Only approved food and water should be consumed. In a suspected biological warfare environment, efforts in monitoring food and water supplies must be increased. Properly treated water and properly cooked food will destroy most biological agents.
   c. Sanitation Measures.
      (1) Maintain high standards of personal hygiene. This will reduce the possibility of catching and spreading infectious diseases.
      (2) Avoid physical fatigue. Physical fatigue lowers the body's resistance to disease. This, of course, is complemented by good physical fitness.
      (3) Stay out of quarantined areas.
      (4) Report sickness promptly. This ensures timely medical treatment and, more importantly, early diagnosis of the disease.
   d. Medical Treatment of Casualties. Once a disease is identified, standard medical treatment commences. This may be in the form of first aid or treatment at a medical facility, depending on the seriousness of the disease. Epidemics of serious diseases may require augmentation of field medical facilities.

7-15. Toxins
Toxins are alleged to have been used in recent conflicts. Witnesses and victims have described the agent as toxic rain (or yellow rain) because it was reported to have been released from aircraft as a yellow powder or liquid that covered the ground, structures, vegetation, and people.

   a. Protective Measures. Individual protective measures normally associated with persistent chemical agents will provide protection against toxins. Measures include the use of the protective mask with hood, and the overgarment ensemble with gloves and overboots (mission-oriented protective posture level-4 [MOPP 4]).

b. Signs/Symptoms. The occurrence of the symptoms from toxins may appear in a period of a few minutes to several hours depending on the particular toxin, the individual susceptibility, and the amount of toxin inhaled, ingested, or deposited on the skin. Symptoms from toxins usually involve the nervous system but are often preceded by less prominent symptoms, such as nausea, vomiting, diarrhea, cramps, or burning distress of the stomach region. Typical neurological symptoms often develop rapidly in severe cases, for example, visual disturbances, inability to swallow, speech difficulty, muscle coordination, and sensory abnormalities (numbness of mouth, throat, or extremities). Yellow rain (mycotoxins) also may have hemorrhagic symptoms which could include any/all of the following:

- Dizziness.
- Severe itching or tingling of the skin.
- Formation of multiple, small, hard blisters.
- Coughing up blood.
- Shock (which could result in death).

c. First Aid Measures. Upon recognition of an attack employing toxins or the onset (start) of symptoms listed above, you must immediately take the following actions:

(1) Step ONE. STOP BREATHING, put on your protective mask with hood, then resume breathing. Next, put on your protective clothing.

(2) Step TWO. Should severe itching of the face become unbearable, quickly—
- Loosen the cap on your canteen.
- Remove your helmet. Take and hold a deep breath and remove your mask.
- While holding your breath, close your eyes and flush your face with generous amounts of water.

⚠ CAUTION
DO NOT rub or scratch your eyes. Try not to let the water run onto your clothing or protective overgarments.

- Put your protective mask back on, seat it properly, clear it, and check it for seal; then resume breathing.
- Put your helmet back on.

> ✐ **NOTE**
> The effectiveness of the M258A1 Skin Decon Kit for biological agent decon is unknown at this time; however, flushing the skin with large amounts of water will reduce the effectiveness of the toxins.

    (3) Step THREE. If vomiting occurs, the mask should be lifted momentarily and drained–while the eyes are closed and the breath is held—and replaced, cleared, and sealed.

  d. Medical Assistance. If you suspect that you have been exposed to toxins, you should seek medical assistance immediately.

7-16. Radiological

There is no direct first aid for radiological casualties. These casualties are treated for their apparent conventional symptoms and injuries.

APPENDIX A

FIRST AID CASE AND KITS, DRESSINGS, AND BANDAGES

A-1. First Aid Case with Field Dressings and Bandages

Every soldier is issued a first aid case (Figure A-1 A) with a field first aid dressing encased in a plastic wrapper (Figure A-13 B). He carries it at all times for his use. The field first aid dressing is a standard sterile (germfree) compress or pad with bandages attached (Figure A-1 C). This dressing is used to cover the wound, to protect against further contamination, and to stop bleeding (pressure dressing). When a soldier administers first aid to another person, he must remember to use the wounded person's dressing; he may need his own later. The soldier must check his first aid case regularly and replace any used or missing dressing. The field first aid dressing may normally be obtained through the medical unit's assigned medical platoon or section.

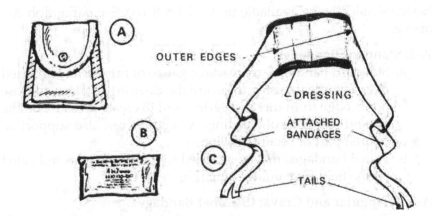

Figure A-1: Field first aid case and dressing

A-2. General Purpose First Aid Kits

General purpose first aid kits listed in paragraph A-3 are also listed in CTA 8-100. These kits are carried on Army vehicles, aircraft, and boats for use by the operators, crew, and passengers. Individuals designated by unit standing operating procedures (SOP) to be responsible for the kits are required to check them regularly and replace all items used, or replace the entire kit when necessary. The general purpose kit and its contents can be obtained through the unit supply system.

✍ NOTE

Periodically check the dressings (for holes or tears in the package) and the medicines (for expiration date) that are in the first aid kits. If necessary, replace defective or outdated items.

A-3. Contents of First Aid Case and Kits

The following items are listed in the Common Table of Allowances (CTA) as indicated below. However, it is necessary to see referenced CTA for stock numbers.

A-4. Dressings

Dressings are sterile pads or compresses used to cover wounds. They usually are made of gauze or cotton wrapped in gauze (Figure A-1 C). In addition to the standard field first aid dressing, other dressings such as sterile gauze compresses and small sterile compresses on adhesive strips may be available under CTA 8-100. See paragraph A-3 above.

A-5. Standard Bandages

a. Standard bandages are made of gauze or muslin and are used over a sterile dressing to secure the dressing in place, to close off its edge from dirt and germs, and to create pressure on the wound and control bleeding. A bandage can also support an injured part or secure a splint.

b. Tailed bandages maybe attached to the dressing as indicated on the field first aid dressing (Figure A-1 C).

A-6. Triangular and Cravat (Swathe) Bandages

a. Triangular and cravat (or swathe) bandages (Figure A-2) are fashioned from a triangular piece of muslin (37 x 37 x 52 inches) provided in the general purpose first aid kit. If it is

CTA	Nomenclature	Unit of Issue	Quantity
a. 50-900	CASE FIELD FIRST AID DRESSING	each	1
8-100	Dressing, first aid field, individual troop, white, 4 by 7 inches	Each	1
b. 8-100	FIRST AID KIT, general purpose (Rigid Case)	Each	1
	Contents:	Each	1
	Case, medical instrument and supply set, plastic, rigid, size A, 7½ inches long by 4½ inches wide by 23/4 inches high		
	Ammonia inhalation solution, aromatic, ampules, 1/3 ml, 10s	Package	1
	Povidone-iodine solution, USP: 10%,½ fl oz, 50s	Box	1/50
	Dressing, first aid, field, individual troop, camou-flaged, 4 by 7 inches	Each	3
	Compress and bandage, camouflaged, 2 by 2 inches, 4s	Package	1
	Bandage, gauze, compressed, camouflaged, 3 inch-es by 6 yards	Each	2
	Bandage, muslin, compressed, camouflaged, 37 by 37 by 52 inches	Each	1
	Gauze, petrolatum, 3 by 36 inches, 3s	Package	1
	Adhesive tape, surgical, 1 inch by 1½ yards, 100s	Package	3/100
	Bandage, adhesive,¾ by 3 inches,	Box	18/300
	Blade, surgical preparation razor, straight, single edge, 5s	Package	1
	First aid kit, eye dressing	Each	1
	Instruction card, artificial respiration, mouth-to-mouth resuscitation (Graphic Training Aid 21-45) (in English)	Each	1
	Instruction sheet, first aid (in English)	Each	1
	Instruction sheet and list of contents (in English)	Each	1
c. 8-100	FIRST AID KIT, general purpose (panel-mounted)	Each	1
	Contents:	Each	1
	Case, medical instrument and supply set, nylon, nonrigid, No. 2, 7½ inches long by 4 3/8 inches wide by 4½ inches high		
In Upper Pocket	Ammonia Inhalation Solution aromatic, ampules, 1/3 ml, 10s	Package	1
	Compress and bandage, camouflaged, 2 by 2 inch-es, 4s	Package	1

continued

CTA	Nomenclature	Unit of Issue	Quantity
	Bandage, muslin, compressed, camouflaged, 37 by 37 by 52 inches	Each	1
	Gauze, petrolatum, 3 by 36 inches, 12s	Package	3/12
	Blade, surgical preparation razor, straight, single edge, 5s	Package	1
In Lower Pocket	Pad, Povidone-Iodine, 100s	Box	10/100
	Dressing, first aid, field, individual troop, camouflaged, 4 by 6 inches	Each	3
	Bandage, gauze, compressed, camouflaged, 3 inches by 6 yards	Each	2
	Adhesive tape, surgical, 1 inch by 1½ yards, 100s	Package	3/100
	Bandage, adhesive, ¾ by 3 inches, 300s	Box	18/300
	First aid kit, eye dressing	Each	1
	Instruction card, artificial respiration, mouth-to-mouth resuscitation (Graphic Training Aid 21-45) (in English)	Each	1
	Instruction sheet, first aid (in English)	Each	1
	Instruction sheet and list of contents (in English)	Each	1

folded into a strip, it is called a cravat. Two safety pins are packaged with each bandage. These bandages are valuable in an emergency since they are easily applied.

b. To improvise a triangular bandage, cut a square of available material, slightly larger than 3 feet by 3 feet, and FOLD it DIAGONALLY. If two bandages are needed, cut the material along the DIAGONAL FOLD.

c. A cravat can be improvised from such common items as T-shirts, other shirts, bed linens, trouser legs, scarfs, or any other item made of pliable and durable material that can be folded, torn, or cut to the desired size.

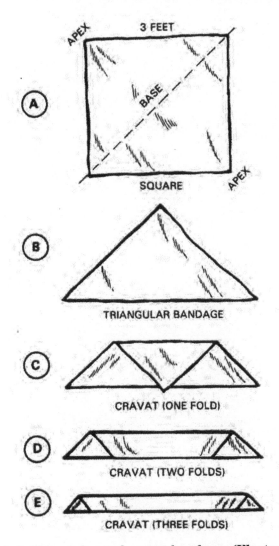

Figure A-2: Triangular and cravat bandages (Illustrated A thru E)

RESCUE AND TRANSPORTATION PROCEDURES

B-1. General

A basic principle of first aid is to treat the casualty before moving him. However, adverse situations or conditions may jeopardize the lives of both the rescuer and the casualty if this is done. It may be necessary first to rescue the casualty before first aid can be effectively or safely given. The life and/or the well-being of the casualty will depend as much upon the manner in which he is rescued and transported as it will upon the treatment he receives. Rescue actions must be done quickly and safely. Careless or rough handling of the casualty during rescue operations can aggravate his injuries and possibly cause death.

B-2. Principles of Rescue Operations

a. When faced with the necessity of rescuing a casualty who is threatened by hostile action, fire, water, or any other immediate hazard, DO NOT take action without first determining the extent of the hazard and your ability to handle the situation. DO NOT become a casualty.

b. The rescuer must evaluate the situation and analyze the factors involved. This evaluation involves three major steps:

- Identify the task.
- Evaluate circumstances of the rescue.
- Plan the action.

B-3. Task (Rescue) Identification

First determine if a rescue attempt is actually needed. It is a waste of time, equipment, and personnel to rescue someone not in need of rescuing. It is also a waste to look for someone who is not lost or needlessly risk the lives of the rescuer(s). In planning a rescue, attempt to obtain the following information:

- Who, what, where, when, why, and how the situation happened?
- How many casualties are involved and the nature of their injuries?
- What is the tactical situation?
- What are the terrain features and the location of the casualties?
- Will there be adequate assistance available to aid in the rescue/evacuation?
- Can treatment be provided at the scene; will the casualties require movement to a safer location?
- What equipment will be required for the rescue operation?
- Will decon procedures and equipment be required for casualties, rescue personnel and rescue equipment?

B-4. Circumstances of the Rescue

a. After identifying the job (task) required, you must relate to the circumstances under which you must work. Do you need additional people, security, medical, or special rescue equipment? Are there circumstances such as mountain rescue or aircraft accidents that may require specialized skills? What is the weather like? Is the terrain hazardous? How much time is available?

b. The time element will sometimes cause a rescuer to compromise planning stages and/or treatment which can be given. A realistic estimate of time available must be made as quickly as possible to determine action time remaining. The key elements are the casualty's condition and the environment.

c. Mass casualties are to be expected on the modern battlefield. All problems or complexities of rescue are now multiplied by the number of casualties encountered. In this case, time becomes the critical element.

B-5. Plan of Action

a. The casualty's ability to endure is of primary importance in estimating the time available. Age and physical condition will

differ from casualty to casualty. Therefore, to determine the
time available, you will have to consider—

- Endurance time of the casualty.
- Type of situation.
- Personnel and/or equipment availability.
- Weather.
- Terrain.

b. In respect to terrain, you must consider altitude and visibil-
   ity. In some cases, the casualty may be of assistance because
   he knows more about the particular terrain or situation than
   you do. Maximum use of secure/reliable trails or roads is
   essential.

c. When taking weather into account, ensure that blankets and/
   or rain gear are available. Even a mild rain can complicate a
   normally simple rescue. In high altitudes and/or extreme cold
   and gusting winds, the time available is critically shortened.

d. High altitudes and gusting winds minimize the ability of
   fixed-wing or rotary wing aircraft to assist in operations. Rotary wing aircraft may be available to remove casualties from
   cliffs or inaccessible sites. These same aircraft can also transport the casualties to a medical treatment facility in a compar-
   atively short time. Aircraft, though vital elements of search,
   rescue or evacuation, cannot be used in all situations. For this
   reason, do not rely entirely on their presence. Reliance on aircraft or specialized equipment is a poor substitute for careful
   planning.

B-6. Mass Casualties

In situations where there are multiple casualties, an orderly rescue
may involve some additional planning. To facilitate a mass casualty
rescue or evacuation, recognize separate stages.

- First Stage. Remove those personnel who are not trapped
  among debris or who can be evacuated easily.
- Second Stage. Remove those personnel who may be trapped
  by debris but require only the equipment on hand and a minimum amount of time.
- Third Stage. Remove the remaining personnel who are
  trapped in extremely difficult or time-consuming situations,
  such asunder large amounts of debris or behind walls.
- Fourth Stage. Remove the dead.

B-7. Proper Handling of Casualties

a. You may have saved the casualty's life through the application of appropriate first aid measures. However, his life can be lost through rough handling or careless transportation procedures. Before you attempt to move the casualty—
- Evaluate the type and extent of his injury.
- Ensure that dressings over wounds are adequately reinforced.
- Ensure that fractured bones are properly immobilized and supported to prevent them from cutting through muscle, blood vessels, and skin. Based upon your evaluation of the type and extent of the casualty's injury and your knowledge of the various manual carries, you must select the best possible method of manual transportation. If the casualty is conscious, tell him how he is to be transported. This will help allay his fear of movement and gain his cooperation and confidence.

b. Buddy aid for chemical agent casualties includes those actions required to prevent an incapacitated casualty from receiving additional injury from the effects of chemical hazards. If a casualty is physically unable to decontaminate himself or administer the proper chemical agent antidote, the casualty's buddy assists him and assumes responsibility for his care. Buddy aid includes—
- Administering the proper chemical agent antidote.
- Decontaminating the incapacitated casualty's exposed skin.
- Ensuring that his protective ensemble remains correctly emplaced.
- Maintaining respiration.
- Controlling bleeding.
- Providing other standard first aid measures.
- Transporting the casualty out of the contaminated area.

B-8. Transportation of Casualties

a. Transportation of the sick and wounded is the responsibility of medical personnel who have been provided special training and equipment. Therefore, unless a good reason for you to transport a casualty arises, wait for some means of medical evacuation to be provided. When the situation is urgent and you are unable to obtain medical assistance or know that no medical evacuation facilities are available, you will have to transport the casualty. For this reason, you must know how to transport him without increasing the seriousness of his condition.

b. Transporting a casualty by litter is safer and more comfortable for him than by manual means; it is also easier for you. Manual transportation, however, may be the only feasible method because of the terrain or the combat situation; or it may be necessary to save a life. In these situations, the casualty should be transferred to a litter as soon as one can be made available or improvised.

B-9. Manual Carries (081-831-1040 and 081-831-1041)

Casualties carried by manual means must be carefully and correctly handled, otherwise their injuries may become more serious or possibly fatal. Situation permitting, evacuation or transport of a casualty should be organized and unhurried. Each movement should be performed as deliberately and gently as possible. Casualties should not be moved before the type and extent of injuries are evaluated and the required emergency medical treatment is given. The exception to this occurs when the situation dictates immediate movement for safety purposes (for example, it may be necessary to remove a casualty from a burning vehicle); that is, the situation dictates that the urgency of casualty movement outweighs the need to administer emergency medical treatment. Manual carries are tiring for the bearer(s) and involve the risk of increasing the severity of the casualty's injury. In some instances, however, they are essential to save the casualty's life. Although manual carries are accomplished by one or two bearers, the two-man carries are used whenever possible. They provide more comfort to the casualty, are less likely to aggravate his injuries, and are also less tiring for the bearers, thus enabling them to carry him farther. The distance a casualty can be carried depends on many factors, such as—

- Strength and endurance of the bearer(s).
- Weight of the casualty.
- Nature of the casualty's injury.
- Obstacles encountered during transport.

a. One-man Carries (081-831-1040).

    (1) Fireman's carry (081-831-1040). The fireman's carry (Figure B-1) is one of the easiest ways for one person to carry another. After an unconscious or disabled casualty has

been properly positioned, he is raised from the ground. An alternate method for raising him from the ground is illustrated (Figure B-1 I). However, it should be used only when the bearer believes it to be safer for the casualty because of the location of his wounds. When the alternate method is used, take care to prevent the casualty's head from snapping back and causing a neck injury. The steps for raising a casualty from the ground for the fireman's carry are also used in other one-man carries.

 KNEEL AT THE CASUALTY'S UNINJURED SIDE, PLACE HIS ARMS ABOVE HIS HEAD AND CROSS HIS ANKLE FARTHER FROM YOU OVER THE ONE CLOSER TO YOU. PLACE ONE OF YOUR HANDS ON THE SHOULDER FARTHER FROM YOU AND YOUR OTHER HAND IN THE AREA OF HIS HIP OR THIGH.

(B)   ROLL HIM TOWARD YOU ONTO HIS ABDOMEN.

Figure B-1: Fireman's carry (Illustrated A thru N)

C AFTER ROLLING THE CASUALTY ONTO HIS ABDOMEN, STRADDLE HIM; THEN PLACE YOUR HANDS UNDER HIS CHEST AND LOCK THEM TOGETHER.

D RAISE/LIFT THE CASUALTY TO HIS KNEES AS YOU MOVE BACKWARD.

E CONTINUE TO MOVE BACKWARD, THUS STRAIGHTENING THE CASUALTY'S LEGS AND LOCKING HIS KNEES.

Figure B-1: *(Continued)*

WALK FORWARD, BRINGING THE CASUALTY TO
A STANDING POSITION BUT TILTED SLIGHTLY
BACKWARD TO PREVENT HIS KNEES FROM
BUCKLING.

(G) AS YOU MAINTAIN CONSTANT SUPPORT OF
THE CASUALTY WITH ONE ARM, FREE YOUR
OTHER ARM, QUICKLY GRASP HIS WRIST,
AND RAISE HIS ARM HIGH.

Figure B-1: *(Continued)*

INSTANTLY PASS YOUR HEAD UNDER HIS RAISED ARM, RELEASING IT AS YOU PASS UNDER IT.

 MOVE SWIFTLY TO FACE THE CASUALTY AND SECURE YOUR ARMS AROUND HIS WAIST. IMMEDIATELY PLACE YOUR FOOT BETWEEN HIS FEET AND SPREAD THEM (APPROXIMATELY 6 TO 8 INCHES APART).

Figure B-1: *(Continued)*

(J) ALTERNATE METHOD OF LIFTING.

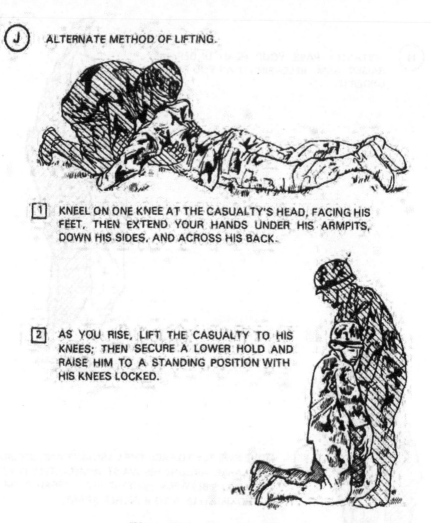

1   KNEEL ON ONE KNEE AT THE CASUALTY'S HEAD, FACING HIS
    FEET, THEN EXTEND YOUR HANDS UNDER HIS ARMPITS,
    DOWN HIS SIDES, AND ACROSS HIS BACK.

2   AS YOU RISE, LIFT THE CASUALTY TO HIS
    KNEES; THEN SECURE A LOWER HOLD AND
    RAISE HIM TO A STANDING POSITION WITH
    HIS KNEES LOCKED.

Figure B-1: *(Continued)*

3 SECURE YOUR ARMS AROUND THE CASUALTY'S WAIST, WITH HIS BODY TILTED SLIGHTLY BACKWARD TO PREVENT HIS KNEES FROM BUCKLING. PLACE YOUR FOOT BETWEEN HIS FEET AND SPREAD THEM (ABOUT 6 TO 8 INCHES APART).

 GRASP THE CASUALTY'S WRIST AND RAISE HIS ARM HIGH OVER YOUR HEAD.

Figure B-1: *(Continued)*

(L) STOOP/BEND DOWN AND PULL THE CASUALTY'S ARM OVER AND DOWN YOUR SHOULDER, THUS BRINGING HIS BODY ACROSS YOUR SHOULDERS. AT THE SAME TIME, PASS YOUR ARM BETWEEN HIS LEGS.

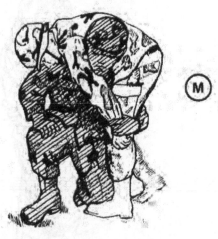

(M) GRASP THE CASUALTY'S WRIST WITH ONE HAND AND PLACE YOUR OTHER HAND ON YOUR KNEE FOR SUPPORT.

Figure B-1: *(Continued)*

RISE WITH THE CASUALTY CORRECTLY POSITIONED.
YOUR OTHER HAND IS FREE FOR USE AS NEEDED.

Figure B-1: *(Continued)*

🖎 NOTE
The alternate method of raising the casualty from the
ground should be used only when the bearer believes
it to be safer for the casualty because of the location of
his wounds. When the alternate method is used, take
care to prevent the casualty's head from snapping
back and causing a neck injury.

---

(2) Support carry (081-831-1040).In the support carry (Figure
B-2), the casualty must be able to walk or at least hop on
one leg, using the bearer as a crutch. This carry can be used
to assist him as far as he is able to walk or hop.

(3) Arms carry (081-831-1040).The arms carry is used when
the casualty is unable to walk. This carry (Figure B-3) is

useful when carrying a casualty for a short distance and when placing him on a litter.

(4) Saddleback carry (081-831-1040). Only a conscious casualty can be transported by the saddleback carry (Figure B-4), because he must be able to hold onto the bearer's neck.

(5) Pack-strap carry (081-831-1040). This carry is used when only a moderate distance will be traveled. In this carry (Figure B-5), the casualty's weight rests high on the bearer's back. To eliminate the possibility of injury to the casualty's arms, the bearer must hold the casualty's arms in a palms-down position.

(6) Pistol-belt carry (081-831-1040). The pistol-belt carry (Figure B-6) is the best one-man carry when the distance to be traveled is long. The casualty is securely supported by a belt upon the shoulders of the bearer. The hands of both the bearer and the casualty are left free for carrying a weapon or equipment, climbing banks, or surmounting obstacles. With his hands free and the casualty secured in place, the bearer is also able to creep through shrubs and under low hanging branches.

RAISE THE CASUALTY TO A STANDING POSITION FROM GROUND AS IN FIREMAN'S CARRY. GRASP THE CASUALTY'S WRIST AND DRAW HIS ARM AROUND YOUR NECK. PLACE YOUR ARM AROUND HIS WAIST.

(THE CASUALTY IS THUS ABLE TO WALK, USING YOU AS A CRUTCH.)

Figure B-2: Support carry

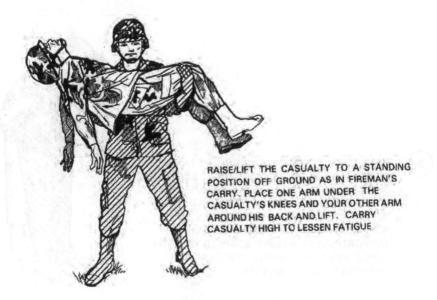

RAISE/LIFT THE CASUALTY TO A STANDING
POSITION OFF GROUND AS IN FIREMAN'S
CARRY. PLACE ONE ARM UNDER THE
CASUALTY'S KNEES AND YOUR OTHER ARM
AROUND HIS BACK AND LIFT. CARRY
CASUALTY HIGH TO LESSEN FATIGUE

Figure B-3: Arms carry

RAISE CASUALTY TO UPRIGHT POSITION AS IN FIREMAN'S
CARRY. SUPPORT CASUALTY BY PLACING AN ARM AROUND
HIS WAIST AND MOVE IN FRONT OF HIM (YOUR BACK TO HIM).
HAVE CASUALTY ENCIRCLE HIS ARMS AROUND YOUR NECK.
STOOP, RAISE HIM UPON YOUR BACK, AND CLASP YOUR
HANDS TOGETHER BENEATH HIS THIGHS IF POSSIBLE.

Figure B-4: Saddleback carry

 LIFT CASUALTY FROM GROUND TO A STANDING
POSITION AS IN FIREMAN'S CARRY. SUPPORTING
THE CASUALTY WITH YOUR ARMS AROUND HIM,
GRASP HIS WRIST CLOSER TO YOU AND PLACE HIS
ARM OVER YOUR HEAD AND ACROSS YOUR
SHOULDER. MOVE IN FRONT OF HIM WHILE
SUPPORTING HIS WEIGHT AGAINST YOUR BACK.
GRASP HIS OTHER WRIST, AND PLACE THIS
ARM OVER YOUR SHOULDER.

BEND FORWARD AND RAISE/HOIST HIM AS
HIGH ON YOUR BACK AS POSSIBLE SO THAT
ALL HIS WEIGHT IS RESTING ON YOUR BACK.

Figure B-5: Pack-strap carry (Illustrated A and B)

(A) LINK TWO PISTOL BELTS (OR THREE, IF NECESSARY) TOGETHER TO FORM A SLING. (IF PISTOL BELTS ARE NOT AVAILABLE FOR USE, OTHER ITEMS, SUCH AS ONE RIFLE SLING, TWO CRAVAT BANDAGES, TWO LITTER STRAPS, OR ANY SUITABLE MATERIAL WHICH WILL NOT CUT OR BIND THE CASUALTY, MAY BE USED.) PLACE THIS SLING UNDER THE CASUALTY'S THIGHS AND LOWER BACK SO THAT A LOOP EXTENDS FROM EACH SIDE.

(B) LIE FACE UP BETWEEN THE CASUALTY'S OUTSTRETCHED LEGS. THRUST YOUR ARMS THROUGH THE LOOPS, GRASP HIS HAND AND TROUSER LEG ON HIS INJURED SIDE.

Figure B-6: Pistol-belt carry (Illustrated A thru F)

(7) Pistol-belt drag (081-831-1040). The pistol-belt drag (Figure B-7) and other drags are generally used for short distances. In this drag the casualty is on his back. The pistol-belt drag is useful in combat. The bearer and the casualty can remain closer to the ground in this drag than in any other.

(8) Neck drag (081-831-1040). The neck drag (Figure B-8) is useful in combat because the bearer can transport

(C) ROLL TOWARD THE CASUALTY'S UNINJURED SIDE ONTO YOUR ABDOMEN, BRINGING HIM ONTO YOUR BACK. ADJUST SLING AS NECESSARY.

(D) RISE TO A KNEELING POSITION. THE BELT WILL HOLD THE CASUALTY IN PLACE.

(E) PLACE ONE HAND ON YOUR KNEE FOR SUPPORT AND RISE TO AN UPRIGHT POSITION.

Figure B-6: *(Continued)*

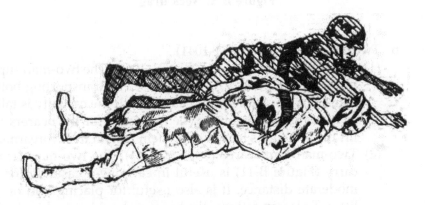

(F) THE CASUALTY IS NOW SUPPORTED ON YOUR SHOULDERS. CARRY THE CASUALTY WITH YOUR HANDS FREE FOR USE IN RIFLE-FIRING, CLIMBING BANKS, OR SURMOUNTING OBSTACLES.

Figure B-6: *(Continued)*

Figure B-7: Pistol-belt drag

the casualty when he creeps behind a low wall or shrubbery, under a vehicle, or through a culvert. This drag is used only if the casualty does not have a broken/fractured arm. In this drag the casualty is on his back. If the casualty is unconscious, protect his head from the ground.

(9) Cradle drop drag (081-831-1040). The cradle drop drag (Figure B-9) is effective in moving a casualty up or down steps. In this drag the casualty is lying down.

TIE THE CASUALTY'S HANDS TOGETHER AT THE WRISTS. IF CASUALTY IS CONSCIOUS, HE MAY CLASP HIS HANDS TOGETHER AROUND YOUR NECK. STRADDLE THE CASUALTY IN A KNEELING FACE-TO-FACE POSITION. LOOP THE CASUALTY'S TIED HANDS OVER/AROUND YOUR NECK. CRAWL FORWARD, LOOKING FORWARD, DRAGGING THE CASUALTY WITH YOU. IF THE CASUALTY IS UNCONSCIOUS, PROTECT HIS HEAD FROM THE GROUND.

Figure B-8: Neck drag

b. Two-man Carries (081-831-1041).
   (1) Two-man support carry (081-831-1041). The two-man support carry (Figure B-10) can be used in transporting both conscious or unconscious casualties. If the casualty is taller than the bearers, it maybe necessary for the bearers to lift the casualty's legs and let them rest on their forearms.
   (2) Two-man arms carry (081-831-1041). The two-man arms carry (Figure B-11) is useful in carrying a casualty for a moderate distance. It is also useful for placing him on a litter. To lessen fatigue, the bearers should carry him high and as close to their chests as possible. In extreme emergencies when there is no time to obtain a board, this manual carry is the safest one for transporting a casualty with a back/neck injury. Use two additional bearers to keep his head and legs in alignment with his body.
   (3) Two-man fore-and-aft carry (081-831-1041). The fore-and-aft carry (Figure B-12) is a most useful two-man carry for transporting a casualty for a long distance. The taller of the two bearers should position himself at the casualty's head. By altering this carry so that both bearers face the casualty, it is also useful for placing him on a litter.

(A) WITH THE CASUALTY LYING ON HIS BACK, KNEEL AT HIS HEAD. THEN SLIDE YOUR HANDS, WITH PALMS UP, UNDER THE CASUALTY'S SHOULDERS AND GET A FIRM HOLD UNDER HIS ARMPITS.

(B) PARTIALLY RISE, SUPPORTING THE CASUALTY'S HEAD ON ONE OF YOUR FOREARMS. (YOU MAY BRING YOUR ELBOWS TOGETHER AND LET THE CASUALTY'S HEAD REST ON BOTH OF YOUR FOREARMS.)

Figure B-9: Cradle drop drag (Illustrated A thru D)

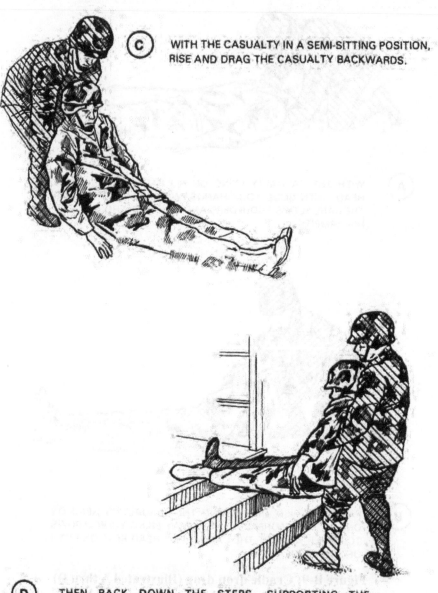

C WITH THE CASUALTY IN A SEMI-SITTING POSITION, RISE AND DRAG THE CASUALTY BACKWARDS.

D THEN BACK DOWN THE STEPS, SUPPORTING THE CASUALTY'S HEAD AND BODY AND LETTING HIS HIPS AND LEGS DROP FROM STEP TO STEP. IF THE CASUALTY NEEDS TO BE MOVED *UP* THE STEPS, THEN YOU SHOULD BACK UP THE STEPS, USING THE SAME PROCEDURE.

Figure B-9: *(Continued)*

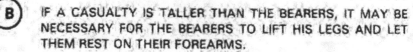

B   IF A CASUALTY IS TALLER THAN THE BEARERS, IT MAY BE
NECESSARY FOR THE BEARERS TO LIFT HIS LEGS AND LET
THEM REST ON THEIR FOREARMS.

Figure B-10: Two-man support carry (Illustrated A and B)

(4) Two-hand seat carry (081-831-1041). The two-hand seat
carry (Figure B-13) is used in carrying a casualty for a
short distance and in placing him on a litter.

(5) Four-hand seat carry (081-831-1041). Only a conscious casualty can be transported with the four-hand seat carry
(Figure B-14) because he must help support himself by
placing his arms around the bearers' shoulders. This carry is especially useful in transporting the casualty with a
head or foot injury and is used when the distance to be
traveled is moderate. It is also useful for placing a casualty
on a litter.

A TWO BEARERS KNEEL AT ONE SIDE OF THE CASUALTY AND PLACE THEIR ARMS BENEATH THE CASUALTY'S BACK (SHOULDERS), WAIST, HIPS, AND KNEES.

B THE BEARERS LIFT THE CASUALTY AS THEY RISE TO THEIR KNEES.

NOTE

Keeping the casualty's body level will prevent unnecessary movement and further injury.

Figure B-11: Two-man arms carry (Illustrated A thru D)

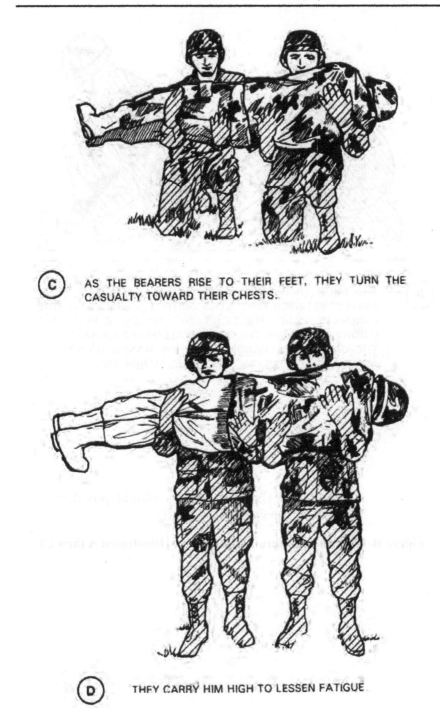

C AS THE BEARERS RISE TO THEIR FEET, THEY TURN THE CASUALTY TOWARD THEIR CHESTS.

D THEY CARRY HIM HIGH TO LESSEN FATIGUE.

Figure B-11: *(Continued)*

A   THE SHORTER BEARER SPREADS THE CASUALTY'S LEGS, KNEELS BETWEEN THE LEGS WITH HIS BACK TO THE CASUALTY, AND POSITIONS HIS HANDS BEHIND THE CASUALTY'S KNEES. THE OTHER (TALLER) BEARER KNEELS AT THE CASUALTY'S HEAD, SLIDES HIS HANDS UNDER THE ARMS AND ACROSS, AND LOCKS HIS HANDS TOGETHER.

NOTE

The taller of the two bearers should position himself at the casualty's head.

Figure B-12: Two-man fore-and-aft carry (Illustrated A thru C)

B THE BEARERS RISE TOGETHER, LIFTING THE CASUALTY.

C ALTERNATE POSITION— FACING CASUALTY.

NOTE

By altering the carry so that both bearers face the casualty, it is also useful for placing him on a litter.

Figure B-12: *(Continued)*

(A) FRONT VIEW

WITH CASUALTY LYING ON HIS BACK, A
BEARER KNEELS ON EACH SIDE OF HIM
AT THE CASUALTY'S HIPS. EACH BEARER
PASSES HIS ARMS UNDER THE
CASUALTY'S THIGHS AND BACK, AND
GRASPS THE OTHER BEARER'S WRISTS.
THE BEARERS RISE, LIFTING THE CASUALTY.

(B) BACK VIEW

Figure B-13: Two-hand seat carry (Illustrated A and B)

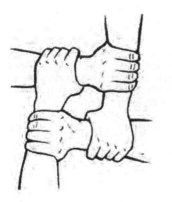

 EACH BEARER GRASPS ONE OF HIS WRISTS AND ONE OF THE OTHER BEARER'S WRISTS, THUS FORMING A PACKSADDLE.

(B) THE TWO BEARERS LOWER THEMSELVES SUFFICIENTLY FOR THE CASUALTY TO SIT ON THE PACKSADDLE; THEN THEY HAVE THE CASUALTY PLACE HIS ARMS AROUND THEIR SHOULDERS FOR SUPPORT BEFORE THEY RISE TO AN UPRIGHT POSITION.

Figure B-14: Four-hand seat carry (Illstrated A and B)

B-10. Improvised Litters (Figures B-15 through B-17) (081-831-1041)
Two men can support or carry a casualty without equipment for only short distances. By using available materials to improvise equipment, the casualty can be transported greater distances by two or more rescuers.

　　a. There are times when a casualty may have to be moved and a standard litter is not available. The distance may be too great for manual carries or the casualty may have an injury, such as a fractured neck, back, hip, or thigh that would be aggravated by manual transportation. In these situations, litters can be

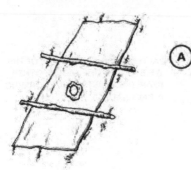

(A) OPEN THE PONCHO AND LAY THE TWO POLES (OR LIMBS) LENGTHWISE ACROSS THE CENTER. REACH IN AND PULL THE HOOD TOWARD YOU AND LAY IT FLAT ON THE PONCHO.

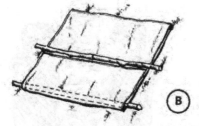

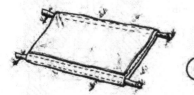

(B) FOLD THE PONCHO OVER THE FIRST POLE.

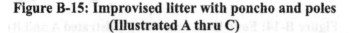

(C) FOLD THE REMAINING FREE EDGES OF THE PONCHO OVER THE SECOND POLE.

**Figure B-15: Improvised litter with poncho and poles
(Illustrated A thru C)**

improvised from certain materials at hand. Improvised litters are emergency measures and must be replaced by standard litters at the first opportunity to ensure the comfort and safety of the casualty.

b. Many different types of litters can be improvised, depending upon the materials available. Satisfactory litters can be made by securing poles inside such items as blankets, ponchos, shelter halves, tarpaulins, jackets, shirts, sacks, bags, and bed tickings (fabric covers of mattresses). Poles can be improvised from strong branches, tent supports, skis, and other like items.

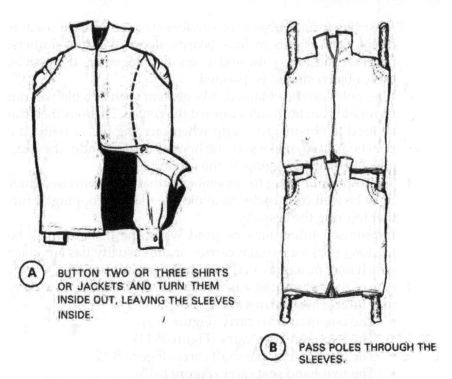

A BUTTON TWO OR THREE SHIRTS OR JACKETS AND TURN THEM INSIDE OUT, LEAVING THE SLEEVES INSIDE.

B PASS POLES THROUGH THE SLEEVES.

Figure B-16: Improvised litter made with poles and jackets (Illustrated A and B)

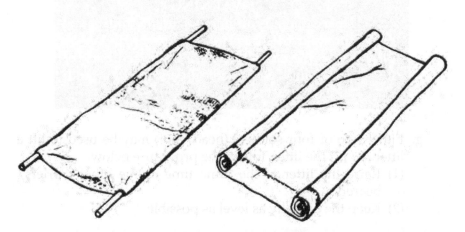

Figure B-17: Improvised litters made by inserting poles through sacks or by rolling blanket

Most flat-surface objects of suitable size can also be used as litters. Such objects include boards, doors, window shutters, benches, ladders, cots, and poles tied together. If possible, these objects should be padded.

c. If no poles can be obtained, a large item such as a blanket can be rolled from both sides toward the center. The rolls then can be used to obtain a firm grip when carrying the casualty. If a poncho is used, make sure the hood is up and under the casualty and is not dragging on the ground.

d. The important thing to remember is that an improvised litter must be well constructed to avoid the risk of dropping or further injuring the casualty.

e. Improvised litters may be used when the distance may be too long (far) for manual carries or the casualty has an injury which may be aggravated by manual transportation.

f. Any of the appropriate carries may be used to place a casualty on a litter. These carries are:
  • The one-man arms carry (Figure B-3).
  • The two-man arms carry (Figure B-11).
  • The two-man fore-and-aft carry (Figure B-12).
  • The two-hand seat carry (Figure B-13).
  • The four-hand seat carry (Figure B-14).

> **WARNING**
>
> Unless there is an immediate life-threatening situation (such as fire, explosion), DO NOT move the casualty with a suspected back or neck injury. Seek medical personnel for guidance on how to transport.

g. Either two or four soldiers (head/foot) may be used to lift a litter. To lift the litter, follow the procedure below.
  (1) Raise the litter at the same time as the other carriers/bearers.
  (2) Keep the casualty as level as possible.

> **✎ NOTE**
>
> Use caution when transporting on a sloping inclines/hills.

APPENDIX C

COMMON PROBLEMS/ CONDITIONS

SECTION I. HEALTH MAINTENANCE

C-1. General

History has often demonstrated that the course of battle is influenced more by the health of the troops than by strategy or tactics. Health is largely a personal responsibility. Correct cleanliness habits, regular exercise, and good nutrition have much control over a person's wellbeing. Good health does not just happen; it comes with conscious effort and good habits. This appendix outlines some basic principles that promote good health.

C-2. Personal Hygiene

a. Because of the close living quarters frequently found in an Army environment, personal hygiene is extremely important. Disease or illness can spread and rapidly affect an entire group.

b. Uncleanliness or disagreeable odors affect the morale of workmates. A daily bath or shower assists in preventing body odor and is necessary to maintain cleanliness. A bath or shower also aids in preventing common skin diseases. Medicated powders and deodorants help keep the skin dry. Special care of the feet is also important. You should wash your feet daily and keep them dry.

C-3. Diarrhea and Dysentery

a. Poor sanitation can contribute to conditions which may result in diarrhea and dysentery (a medical term applied to a number of intestinal disorders characterized by stomach pain and diarrhea with passage of mucus and blood). Medical personnel can advise regarding the cause and degree of illness.

Remember, however, that intestinal diseases are usually spread through contact with infectious organisms which can be spread in human waste, by flies and other insects, or in improperly prepared or disinfected food and water supplies.

b. Keep in mind the following principles that will assist you in preventing diarrhea and/or dysentery.

(1) Fill your canteen with treated water at every chance. When treated water is not available you must disinfect the water in your canteen by boiling it or using either iodine tablets or chlorine ampules. Iodine tablets or chlorine ampules can be obtained through your unit supply channels or field sanitation team.

(a) To treat (disinfect) water by boiling, bring water to a rolling boil in your canteen cup for 5 to 10 minutes. In an emergency, boiling water for even 15 seconds will help. Allow the water to cool before drinking.

(b) To treat water with iodine—

• Remove the cap from your canteen and fill the canteen with the cleanest water available.

• Put one tablet in clear water or two tablets in very cold or cloudy water. Double amounts if using a two quart canteen.

• Replace the cap, wait 5 minutes, then shake the canteen. Loosen the cap and tip the canteen over to allow leakage around the canteen threads. Tighten the cap and wait an additional 25minutes before drinking.

(c) To treat water with chlorine—

• Remove the cap from your canteen and fill your canteen with the cleanest water available.

• Mix one ampule of chlorine with one-half canteen cup of water, stir the mixture with a mess kit spoon until the contents are dissolved. Take care not to cut your hands when breaking open the glass ampule.

• Pour one canteen capful of the chlorine solution into your one quart canteen of water.

• Replace the cap and shake the canteen. Loosen the cap and tip the canteen over to allow leakage around the threads. Tighten the cap and wait 30 minutes before drinking.

(2) DO NOT buy food, drinks, or ice from civilian vendors unless approved by medical personnel.
(3) Wash your hands for at least 30 seconds after using the latrine or before touching food.
(4) Wash your mess kit in a mess kit laundry or with treated water.
(5) Food waste should be disposed of properly (covered container, plastic bags or buried) to prevent flies from using it as a breeding area.

C-4. Dental Hygiene

    a. Care of the mouth and teeth by daily use of a toothbrush and dental floss after meals is essential. This care may prevent gum disease, infection, and tooth decay.

    b. One of the major causes of tooth decay and gum disease is plaque. Plaque is an almost invisible film of decomposed food particles and millions of living bacteria. To prevent dental diseases, you must effectively remove this destructive plaque.

C-5. Drug (Substance) Abuse

    a. Drug abuse is a serious problem in the military. It affects combat readiness, job performance, and the health of military personnel and their families. More specifically, drug abuse affects the individual. It costs millions of dollars in lost time and productivity.

    b. The reasons for drug abuse are as different as the people who abuse the use of them. Generally, people seem to take drugs to change the way they feel. They may want to feel better or to feel happier. They may want to escape from pain, stress, or frustration. Some may want to forget. Some may want to be accepted or to be sociable. Some people take drugs to escape boredom; some take drugs because they are curious. Peer pressure can also be a very strong reason to use drugs.

    c. People often feel better about themselves when they use drugs or alcohol, but the effects do not last. Drugs never solve problems; they just postpone or compound them. People who abuse alcohol or drugs to solve one problem run the risk of continued drug use that creates new problems and makes old problems worse.

d. Drug abuse is very serious and may cause serious health problems. Drug abuse may cause mental incapacitation and even cause death.

C-6. Sexually Transmitted Diseases

Sexually transmitted diseases (STD) formerly known as venereal diseases are caused by organisms normally transmitted through sexual intercourse. Individuals should use a prophylactic (condom) during sexual intercourse unless they have sex only within marriage or with one, steady non infected person of the opposite sex. Another good habit is to wash the sexual parts and urinate immediately after sexual intercourse, Some serious STDs include nonspecific urethritis (chlamydia), gonorrhea, syphilis, and Hepatitis B and the Acquired Immunodeficiency Syndrome (AIDS). Prevention of one type of STD through responsible sex, protects both partners from all STD. Seek the best medical attention if any discharge or blisters are found on your sexual parts.

a. Acquired Immunodeficiency Syndrome (AIDS). AIDS is the end disease stage of the HIV infection. The HIV infection is contagious, but it cannot be spread in the same manner as a common cold, measles, or chicken pox. AIDS is contagious, however, in the same way that sexually transmitted diseases, such as syphilis and gonorrhea, are contagious. AIDS can also be spread through the sharing of intravenous drug needles and syringes used for injecting illicit drugs.

b. High Risk Group. Today those practicing high risk behavior who become infected with the AIDS virus are found mainly among homosexual and bisexual persons and intravenous drug users. Heterosexual transmission is expected to account for an increasing proportion of those who become infected with the AIDS virus in the future.

(1) AIDS caused by virus. The letters A-I-D-S stand for Acquired Immunodeficiency Syndrome. When a person is sick with AIDS, he is in the final stages of a series of health problems caused by a virus (germ) that can be passed from one person to another chiefly during sexual contact or through the sharing of intravenous drug needles and syringes used for "shooting" drugs. Scientists have named the AIDS virus "HIV." The HIV attacks a person's immune system and damages his ability to fight other disease.

Without a functioning immune system to ward off other germs, he now becomes vulnerable to becoming infected by bacteria, protozoa, fungi, and other viruses and malignancies, which may cause life-threatening illness, such as pneumonia, meningitis, and cancer.

(2) No known cure. There is presently no cure for AIDS. There is presently no vaccine to prevent AIDS.

(3) Virus invades blood stream. When the AIDS virus enters the blood stream, it begins to attack certain white blood cells (T-Lymphocytes). Substances called antibodies are produced by the body. These antibodies can be detected in the blood by a simple test, usually two weeks to three months after infection. Even before the antibody test is positive, the victim can pass the virus to others.

(4) Signs and Symptoms.

- Some people remain apparently well after infection with the AIDS virus. They may have no physically apparent symptom of illness. However, if proper precautions are not used with sexual contacts and/or intravenous drug use, these infected individuals can spread the virus to others.

- The AIDS virus may also attack the nervous system and cause delayed damage to the brain. This damage may take years to develop and the symptoms may show up as memory loss, indifference, loss of coordination, partial paralysis, or mental disorder. These symptoms may occur alone, or with other symptoms mentioned earlier.

(5) AIDS: the present situation. The number of people estimated to be infected with the AIDS virus in the United States is over 1.5 million as of April 1988. In certain parts of central Africa 50% of the sexually active population is infected with HIV. The number of persons known to have AIDS in the United States to date is over 55,000; of these, about half have died of the disease. There is no cure. The others will soon die from their disease. Most scientists predict that all HIV infected persons will develop AIDS sooner or later, if they don't die of other causes first.

(6) Sex between men. Men who have sexual relations with other men are especially at risk. About 70% of AIDS victims throughout the country are male homosexuals and

bisexuals. This percentage probably will decline as heterosexual transmission increases. Infection results from a sexual relationship with an infected person.

(7) Multiple partners. The risk of infection increases according to the number of sexual partners one has, male or female. The more partners you have, the greater the risk of becoming infected with the AIDS virus.

(8) How exposed. Although the AIDS virus is found in several body fluids, a person acquires the virus during sexual contact with an infected person's blood or semen and possibly vaginal secretions. The virus then enters a person's blood stream through their rectum, vagina or penis. Small (unseen by the naked eye) tears in the surface lining of the vagina or rectum may occur during insertion of the penis, fingers, or other objects, thus opening an avenue for entrance of the virus directly into the blood stream.

(9) Prevention of sexual transmission—know your partner. Couples who maintain mutually faithful monogamous relationships (only one continuing sexual partner) are protected from AIDS through sexual transmission. If you have been faithful for at least five years and your partner has been faithful too, neither of you is at risk.

(10) Mother can infect newborn. If a woman is infected with the AIDS virus and becomes pregnant, she has about a 50% chance of passing the AIDS virus to her unborn child.

(11) Summary. AIDS affects certain groups of the population. Homosexual and bisexual persons who have had sexual contact with other homosexual or bisexual persons as well as those who "shoot" street drugs are at greatest risk of exposure, infections and eventual death. Sexual partners of these high risk individuals are at risk, as well as any children born to women who carry the virus. Heterosexual persons are increasingly at risk.

(12) Donating blood. Donating blood is not risky at all. You cannot get AIDS by donating blood.

(13) Receiving blood. High risk persons and every blood donation is now tested for the presence of antibodies to the AIDS virus. Blood that shows exposure to the AIDS virus by the presence of antibodies is not used either for transfusion or for the manufacture of blood products. Blood banks are as safe as current technology can make them. Because

antibodies do not form immediately after exposure to the virus, a newly infected person may unknowingly donate blood after becoming infected but before his antibody test becomes positive.

(14) Testing of military personnel. You may wonder why the Department of Defense currently tests its uniformed services personnel for presence of the AIDS virus antibody. The military feels this procedure is necessary because the uniformed services act as their own blood bank in a combat situation. They also need to protect new recruits (who unknowingly may be AIDS virus carriers) from receiving live virus vaccines. HIV antibody positive soldiers may not be assigned overseas (includes Alaska and Hawaii). They must be rechecked every six months to determine if the disease has become worse. If the disease has progressed, they are discharged from the Army (policy per AR 600-110). This regulation requires that all soldiers receive annual education classes on AIDS.

SECTION II. FIRST AID FOR COMMON PROBLEMS

C-7. Heat Rash (or Prickly Heat)

a. Description. Heat rash is a skin rash caused by the blockage of the sweat glands because of hot, humid weather or because of fever. It appears as a rash of patches of tiny reddish pinpoints that itch.

b. First Aid. Wear clothing that is light and loose and/or uncover the affected area. Use skin powders or lotion.

Figure C-1: Poison ivy

Figure C-2: Western poison oak

Figure C-3: Poison sumac

C-8. Contact Poisoning (Skin Rashes)

a General.

   (1) Poison Ivy grows as a small plant (vine or shrub) and has three glossy leaflets (Figure C-1).

   (2) Poison Oak grows in shrub or vine form; and has clusters of three leaflets with wavy edges (Figure C-2).

   (3) Poison Sumac grows as a shrub or small tree. Leaflets grow opposite each other with one at tip (Figure C-3).

b. Signs/Symptoms.

- Redness.
- Swelling.
- Itching.
- Rashes or blisters.
- Burning sensation.
- General headaches and fever.

✍ **NOTE**

Secondary infection may occur when blisters break.

---

c. First Aid.
   (1) Expose the affected area: remove clothing and jewelry.
   (2) Cleanse affected area with soap and water.
   (3) Apply rubbing alcohol, if available, to the affected areas.
   (4) Apply calamine lotion (helps relieve itching and burning).
   (5) Avoid dressing the affected area.
   (6) Seek medical help, evacuate if necessary. (If rash is severe, or on face or genitals, seek medical help.)

C-9. Care of the Feet

Proper foot care is essential for all soldiers in order to maintain their optimal health and physical fitness. To reduce the possibilities of serious foot trouble, observe the following rules:

   a. Foot hygiene is important. Wash and dry feet thoroughly, especially between the toes. Soldiers who perspire freely should apply powder lightly and evenly twice a day.
   b. Properly fitted shoes/boots should be the only ones issued. There should be no binding or pressure spots.
   c. Clean, properly fitting socks should be changed and washed daily. Avoid socks with holes or poorly darned areas; they may cause blisters.
   d. Attend promptly to common medical problems such as blisters, ingrown toenails, and fungus infections (like athlete's foot).
   e. Foot marches are a severe test for the feet. Use only properly fitted footgear and socks. Footgear should be completely broken-in. DO NOT break-in new footgear on a long march. Any blisters, sores, and so forth, should be treated promptly. Keep the feet as dry as possible on the march; carry extra socks and change if feet get wet (socks can be dried by putting them under your shirt, around your waist or hanging on a rack). Inspect feet during rest breaks. Bring persistent complaints to the attention of medical personnel.

* C-10. Blisters

Blisters are a common problem caused by friction. They may appear on such areas as the toes, heels, or the palm of the hand (anywhere friction may occur). Unless treated promptly and correctly, they may become infected. PREVENTION is the best solution to AVOID blisters and subsequent infection. For example, ensure boots are prepared properly for a good fit, whenever possible always keep feet clean and dry; and, wear clean socks that also fit properly. Gloves should be worn whenever extensive manual work is done.

> ✍ **NOTE**
>
> Keep blisters clean. Care should be taken to keep the feet as clean as possible at all times. Use soap and water for cleansing. Painful blisters and/or signs of infection, such as redness, throbbing, drainage, and so forth, are reasons for seeking medical treatment. Seek medical treatment only from qualified medical personnel.

APPENDIX D

DIGITAL PRESSURE

APPLY DIGITAL PRESSURE

Digital pressure (also often called "pressure points") is an alternate method to control bleeding. This method uses pressure from the fingers, thumbs, or hands to press at the site or point where a main artery supplying the wounded area lies near the skin surface or over

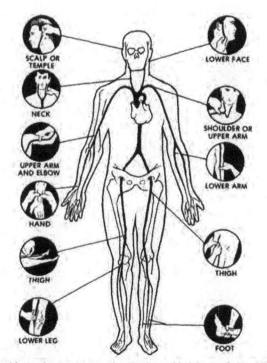

If blood is spurting from wound (artery), press at the point or site where main artery supplying the wounded area lies near skin surface or over bone as shown. This pressure shuts off or slows down the flow of blood from the heart to the wound until a pressure dressing can be unwrapped and applied. You will know you have located the artery when you feel a pulse.

Figure D-1: Digital pressure (pressure with fingers, thumbs or hands)

bone (Figure D-1). This pressure may help shut off or slow down the flow of blood from the heart to the wound and is used in combination with direct pressure and elevation. It may help in instances where bleeding is not easily controlled, where a pressure dressing has not yet been applied, or where pressure dressings are not readily available.

APPENDIX E

DECONTAMINATION PROCEDURES

APPLY DIGITAL PRESSURE

E-1. Protective Measures and Handling of Casualties

a. Depending on the theater of operations, guidance issued may dictate the assumption of a minimum mission-oriented protective posture (MOPP) level. However, a full protective posture (MOPP 4) level will be assumed immediately when the alarm or command is given. (MOPP 4 level consists of wearing the protective over garment, mask, hood, gloves, and overboots.) If individuals find themselves alone without adequate guidance, they should mask and assume the MOPP 4 level under any of the following conditions.

    (1) Their position is hit by a concentration of artillery, mortar, rocket fire, or by aircraft bombs if chemical agents have been used or the threat of their use is significant.

    (2) Their position is under attack by aircraft spray.

    (3) Smoke or mist of an unknown source is present or approaching.

    (4) A suspicious odor or a suspicious liquid is present.

    (5) A toxic chemical or biological attack is suspected.

    (6) They are entering an area known to be or suspected of being contaminated with a toxic chemical or biological agent.

    (7) During any motor march, once chemical warfare has been initiated.

    (8) When casualties are being received from an area where chemical agents have reportedly been used.

    (9) They have one or more of the following signs/symptoms:

        (a) An unexplained sudden runny nose.

        (b) A feeling of choking or tightness in the chest or throat.

   (c) Blurring of vision and difficulty in focusing the eyes on close objects.

   (d) Irritation of the eyes (could be caused by the presence of several toxic chemical agents).

   (e) Unexplained difficulty in breathing or increased rate of breathing.

   (f) Sudden feeling of depression.

   (g) Dread, anxiety, restlessness.

   (h) Dizziness or light-headedness.

   (i) Slurred speech.

  (10) Unexplained laughter or unusual behavior noted in others.

  (11) Buddies suddenly collapsing without evident cause.

 b. Stop breathing don the protective mask, seat it properly, clear it, and check it for seal; then resume breathing. The mask should be worn until unmasking procedures indicate no chemical agent is in the air and the "all clear" signal is given. If vomiting occurs, the mask should be lifted momentarily and drained—while the eyes are closed and the breath is held—and replaced, cleared, and sealed.

 c. Casualties contaminated with a chemical agent may endanger unprotected personnel. Handlers of these casualties must wear a protective mask, protective gloves, and chemical protective clothing until the casualty's contaminated clothing has been removed. The battalion aid station should be established upwind from the most heavily contaminated areas, if it is expected that troops will remain in the area six hours or more. Collective protective shelters must be used to adequately manage casualties on the integrated battlefield. Casualties must be undressed and decontaminated, as required, in an area equipped for the removal of contaminated clothing and equipment prior to entering collective protection. Contaminated clothing and equipment should be placed in airtight containers or plastic bags, if available, or removed to a designated dump site downwind from the aid station.

F-2. Personal Decontamination

Following contamination of the skin or eyes with vesicants (mustards, lewisite, and so forth) or nerve agents, personal decontamination must be carried out immediately. This is because chemical agents are effective at very small concentrations and within a very few minutes after exposure, decontamination is marginally effective.

Decontamination consists of either removal and/or neutralization of the agent. Decontamination after absorption occurs may serve little or no purpose. Soldiers will decontaminate themselves unless they are incapacitated. For soldiers who cannot decontaminate themselves, the nearest able person should assist them as the situation permits.

> ✍ **NOTE**
>
> In a cyanide only environment, there would be no need for decontamination.

a. Eyes. Following contamination of the eyes with any chemical agent, the agent must be removed instantly. In most cases, identity of the agent will not be known immediately. Individuals who suspect contamination of their eyes or face must quickly obtain overhead shelter to protect themselves while performing the following decontamination process:

(1) Remove and open your canteen.

(2) Take a deep breath and hold it.

(3) Remove the mask.

(4) Flush or irrigate the eye, or eyes, immediately with large amounts of water. To flush the eyes with water from a canteen (or other container of uncontaminated water), tilt the head to one side, open the eyelids as wide as possible, and pour water slowly into the eye so that it will run off the side of the face to avoid spreading the contamination. This irrigation must be carried out despite the presence of toxic vapors in the atmosphere. Hold your breath and keep your mouth closed during this procedure to prevent contamination and absorption through the mucous membranes. Chemical residue flushed from the eyes should be neutralized along the flush path.

WARNING

DO NOT use the fingers or gloved hands for holding the eyelids apart. Instead, open the eyes as wide as possible and pour the water as indicated above.

(5) Replace, clear, and check your mask. Then resume breathing.

(6) If contamination was picked up while flushing the eyes, then decontaminate the face. Follow procedure outlined in paragraph b (2) *(a)* through *(ae)* below.

b. Skin (Hands, Face, Neck, Ears, and Other Exposed Areas). The M258A1 Skin Decontamination Kit (Figure F-1) is provided individuals for performing emergency decontamination of their skin (and selected small equipment, such as the protective gloves, mask, hood, and individual weapon).

(1) Description of the M258A1 kit. The M258A1 kit measures 1 3/4 by 2 3/4 by 4 inches and weighs 0.2 pounds. Each kit contains six packets: three DECON-1 packets and three DECON-2packets. DECON-1 packet contains a pad pre-moistened with hydroxyethane 72%, phenol 10%, sodium hydroxide 5%, and ammonia 0.2%, and the remainder water. DECON-2 packet contains a pad impregnated with chloramine B and sealed glass ampules filled with hydroxyethane 45%, zinc chloride 5%, and the remainder water. The case fits into the pocket on the outside rear of the M17 series protective mask carrier or in an inside

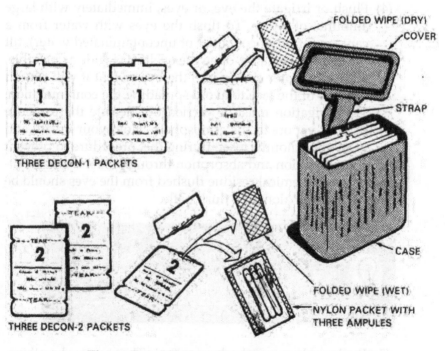

Figure E-1: M258A1 Skin Decontamination Kit

pocket of the carrier for the M24 and M25 series protective mask. The case can also be attached to the web belt or on the D ring of the protective mask carrier.

(2) Use of the M258A1 kit. It should be noted that the procedures outlined in paragraphs *(a)* through *(ae)* below were not intended to replace or supplant those contained in STP 21-1-SMCT but, rather, to expand on the doctrine of skin decontamination.

WARNING

The ingredients of the DECON-1 and DECON-2 packets of the M258A1 kit are, poisonous and caustic and can permanently damage the eyes. KEEP PADS OUT OF THE EYES, MOUTH, AND OPEN WOUNDS. Use water to wash the toxic agent out of the eyes or wounds, except in the case of mustard, Mustard may be removed by thorough immediate wiping.

WARNING

The complete decon (WIPES 1 and 2) of the face must be done as quickly as possible–3 minutes or less.

WARNING

DO NOT attempt to decontaminate the face or neck before putting on a protective mask.

✍ NOTE

Use the buddy system to decontaminate exposed skin areas you cannot reach.

✍ NOTE

Blisters caused by blister agents are actually burns and should be treated as such. Blisters which have ruptured are treated as open wounds.

(a) Put on the protective mask (if not already on).
(b) Seek overhead cover or use a poncho for protection against further contamination.
(c) Remove the M258A1 kit. Open the kit and remove one DECON-1 WIPE packet by its tab.
(d) Fold the packet on the solid line marked BEND, then unfold it.
(e) Tear open the packet quickly at the notch, and remove the wipe and fully open it.
(f) Wipe your hands.

✍ NOTE

If you have a chemical agent on your face, do steps *(g)* through *(t)*. If you do not have an agent on your face, do step *(m)*, continue to decon other areas of contaminated skin, then go to step *(n)*.

✍ NOTE

You must hold your breath while doing steps *(g)* through *(l)*. If you need to breathe before you finish, reseal your mask, clear it and check it, then continue.

(g) Hold your breath, close your eyes, and lift the hood and mask from your chin.
(h) Scrub up and down from ear to ear.
　1. Start at an ear.
　2. Scrub across the face to the corner of the nose.
　3. Scrub an extra stroke at the corner of the nose.
　4. Scrub across the nose and tip of the nose to the corner of the nose.
　5. Scrub an extra stroke at the corner of the nose.
　6. Scrub across the face to the other ear.
(i) Scrub up and down from the ear to the end of the jawbone.
　1. Begin where step (h) ended.
　2. Scrub across the cheek to the corner of the mouth.
　3. Scrub an extra stroke at the corner of the mouth.
　4. Scrub across the closed mouth to the center of the upper lip.
　5. Scrub an extra stroke above the upper lip.

6. Scrub across the closed mouth to the corner of the mouth.

7. Scrub an extra stroke at the corner of the mouth.

8. Scrub across the cheek to the end of the jawbone.

(j) Scrub up and down from one end of the jawbone to the other end of the jawbone.

1. Begin where step (i) ended.

2. Scrub across and under the jaw to the chin, cupping the chin.

3. Scrub an extra stroke at the cleft of the chin.

4. Scrub across and under the jaw to the end of the jawbone.

(k) Quickly wipe the inside of the mask which touches the face.

(l) Reseal, clear, and check the mask. Resume breathing.

(m) Using the same DECON-1 WIPE, scrub the neck and the ears.

(n) Rewipe the hands.

(o) Drop the wipe to the ground.

(p) Remove one DECON-2 WIPE packet, and crush the encased glass ampules between the thumb and fingers. DO NOTKNEAD.

(q) Fold the packet on the solid line marked CRUSH AND BEND, then unfold it.

(r) Tear open the packet quickly at the notch and remove the wipe.

(s) Fully open the wipe. Let the encased crushed glass ampules fall to the ground.

(t) Wipe your hands.

✍ NOTE

If you have an agent on your face, do steps (u) through (ae). If you do not have an agent on your face, do step (aa), continue to decon other areas of contaminated skin, then go to step (ab).

✍ NOTE

You must hold your breath while doing steps (u) through (z). If you need to breathe before you finish, reseal your mask, clear it and check it, then continue.

(u) Hold your breath, close your eyes, and lift the hood and mask away from your chin.

(v) Scrub up and down from ear to ear.
1. Start at an ear.
2. Scrub across the face to the corner of the nose.
3. Scrub an extra stroke at the corner of the nose.
4. Scrub across the nose and tip of the nose to the corner of the nose.
5. Scrub an extra stroke at the corner of the nose.
6. Scrub across the face to the other ear.

(w) Scrub up and down from the ear to the end of the jawbone.
1. Begin where step (v) ended.
2. Scrub across the cheek to the corner of the mouth.
3. Scrub an extra stroke at the corner of the mouth.
4. Scrub across the closed mouth to the center of the upper lip.
5. Scrub an extra stroke above the upper lip.
6. Scrub across the closed mouth to the corner of the mouth.
7. Scrub an extra stroke at the corner of the mouth.
8. Scrub across the cheek to the end of the jawbone.

(x) Scrub up and down from one end of the jawbone to the other end of the jawbone.
1. Begin where step (w) ended.
2. Scrub across and under the jaw to the chin, cupping the chin.
3. Scrub an extra stroke at the cleft of the chin.
4. Scrub across and under the jaw to the end of the jawbone.

(y) Quickly wipe the inside of the mask which touches the face.

(z) Reseal, clear, and check the mask. Resume breathing.

(aa) Using the same DECON-2 WIPE, scrub the neck and ears.

(ab) Rewipe the hands.

(ac) Drop the wipe to the ground.

(ad) Put on the protective gloves and any other protective clothing, as appropriate. Fasten the hood straps and neck cord.

(ae) Bury the decontaminating packet and other items dropped on the ground, if circumstances permit.

c. Clothing and Equipment. Although the M258A1 may be used for decontamination of selected items of individual clothing and equipment (for example, the soldier's individual weapon), there is insufficient capability to do more than emergency spot decontamination. The M258A1 is not used to decontaminate the protective overgarment. The protective overgarment does not require immediate decontamination since the charcoal layer is a decontaminating device; however, it must be exchanged. The Individual Equipment Decontamination Kit (DKIE), M280 (similar in configuration to the M258A1), is used to decontaminate equipment such as the weapon, helmet, and other gear that is carried by the individual.

E-3. Casualty Decontamination

Contaminated casualties entering the medical treatment system are decontaminated through a decentralized process. This is initially started through self-aid and buddy aid procedures. Later, units should further decontaminate the casualty before evacuation. Casualty decontamination stations are established at the field medical treatment facility to further decontaminate these individuals (clothing removal and spot decontamination, as required) prior to treatment and evacuation. These stations are manned by nonmedical members of the supported unit under supervision of medical personnel. There are insufficient medical personnel to both decontaminate and treat casualties. The medical personnel must be available for treatment of the casualties during and after decontamination by nonmedical personnel. Decontamination is accomplished as quickly as possible to facilitate medical treatment, prevent the casualty from absorbing additional agent, and reduce the spread of chemical contamination.

Notes